SCOTNOTES
Number 45

Exploring Scottish Place-Names

A' Fuasgladh Ainmean-Àite na h-Alba

John Hodgart

Association for Scottish Literature 2024

Published by
Association for Scottish Literature
Scottish Literature
7 University Gardens
University of Glasgow
Glasgow G12 8QH
www.asls.org.uk

ASL is a registered charity no. SC006535

First published 2024

Text © John Hodgart

All rights reserved. No part of this book may be reproduced, stored in a retrieval system, or transmitted in any form or means, electronic, mechanical, photocopying, recording or otherwise, without the prior permission of the Association for Scottish Literature.

A CIP catalogue for this title
is available from the British Library

ISBN 978-1-906871-60-7

ASL acknowledges the support of the
Scottish Government towards the publication of this book

CONTENTS

1.	Introduction	1
2.	Why Study Place-Names?	10
3.	A Guide for Teachers and Students	12
	The Spelling is Atroshus!	
	But Listen to the Locals	
4.	The Structure of Place-Names – the Building-Blocks	21
	Other Categories or Classifications	
5.	Index of Key Names and Their Linguistic Origins	27
6.	Pre-Celtic / Early Celtic	35
	Pictish – the Language of Calgacus	
7.	Brittonic / Brythonic or Cumbric / Cymric	42
8.	Gaelic – the Language of the First 'Scots'	55
9.	Norse – the Language of the Vikings	96
10.	Old English / Anglic / Scots	110
11.	Other Languages	128
12.	Modern Names – Social, Political and Cultural Issues	133
13.	Work and Place-Names	142
14.	Scottish Names Worldwide	146
15.	Conclusion	150
16.	Bibliography	151
17.	Online Resources	153

SCOTNOTES

Study guides to major Scottish writers and literary texts

Produced by the Education Committee
of the Association for Scottish Literature

Series Editors
Lorna Borrowman Smith
Ronald Renton

Editorial Board
Dr Ronald Renton
(Convener, Education Committee, ASL)
Laurence Cavanagh
Professor John Corbett
Dr Emma Dymock
Dr Maureen Farrell
Dr Morna Fleming
John Hodgart
Bob Hume
Ann MacKinnon
Dr Maria Marchidanu
Professor Alan Riach
Dr Gillian Sargent
Dr Cheryl Simpson
Lorna Borrowman Smith
Andrew Young

THE ASSOCIATION FOR SCOTTISH LITERATURE aims to promote the study, teaching and writing of Scottish literature, and to further the study of the languages of Scotland.

To these ends, the ASL publishes works of Scottish literature; literary criticism and in-depth reviews of Scottish books in *Scottish Literary Review*; and scholarly studies of language in *Scottish Language*. It also publishes *New Writing Scotland*, an annual anthology of new poetry, drama and short fiction, in Scots, English and Gaelic. All these publications are available as a single 'package', in return for an annual subscription.

ASL also produces a range of teaching materials covering Scottish language and literature for use in schools.

Enquiries should be sent to:

>ASL
>Scottish Literature
>7 University Gardens
>University of Glasgow
>Glasgow G12 8QH
>
>Tel/fax +44 (0)141 330 5309
>e-mail **office@asls.org.uk**
>or visit our website at **www.asls.org.uk**

Acknowledgements

This Scotnote would never have been completed without the advice, support, meticulous editing and generous friendship of Ronnie Renton over many years. I must also thank my wife Mhairi for her amazing curiosity and memory, especially in presenting me with many 'whitaboot?' lists of Scottish place-names. I also need to thank Ian MacDonald for casting his sharp eye over it, especially the Gaelic grammar and spelling, and also Alasdair Whyte for his helpful advice on the Gaelic section.

Abbreviations

Br	Brittonic
OE	Old English
ON	Old Norse
P	Pictish
S	Scots
gen.	genitive
nom.	nominative
pl	plural
pron.	pronounced

1. INTRODUCTION

This Scotnote is different from others in our series as it does not deal with a text or an author, but provides a guide to Scottish place-names and their linguistic origins. Its aims are firstly to provide an introduction to Scotland's linguistic heritage and secondly to offer a basic understanding of some key place-name elements in our landscape. Unlike place-names guides or dictionaries, it does not just give an alphabetical list of names, but introduces readers to key words and the basic tools to help them work out meanings for themselves.

Hopefully it will appeal to anyone with an interest in Scotland's place-names, though it is primarily intended as a guide for teachers, senior pupils and students. It provides a short guide to the history of each of the languages from which our place-names originate, followed by a concise outline of some key or common place-name elements, with examples in each of these languages, plus sections (Gaelic, Norse and Scots) on farming, flora and fauna. The social, political, cultural and industrial dimensions behind modern names are also discussed at the end, plus a brief look at Scottish names worldwide.

A classroom guide is also available in the Schools section of the ASL website as a separate Teaching Note which offers a short introduction to place-name elements, plus a table of common elements in each of those languages, followed by some suggestions and activities for studying place-names and for investigation of the local area.

The genesis of this Scotnote lies in a short teaching unit which Ronnie Renton produced many years ago, with some help from the late Willie Smith, for the Jordanhill Scottish Literature and Language Project which Ronnie and I were involved with when we were eager young teachers in the late 1970s. I tried some activities on place-names in my own department over the years, though I never succeeded in interesting my Social Subjects colleagues, at least not enough to try a cross-curricular project on place-names, in spite of the fact that I usually found it ran in all directions, sometimes like a burn in spate!

I must stress from the start that I am not an expert on this topic, only an enthusiastic scholar, who, in his teaching days, tried to interest weans in place-names, starting with the name of their school, Garnock Academy, in North Ayrshire, an the three toons o the Garnock or Gaurnock Valley (Dalry, Kilbirnie and Beith). An rarely did they ken whit these names meant!

However, my own fascination with place-names goes back a long way, possibly to a time I wasn't even aware that names had meanings, but a time when the sound of names rang or echoed in my mind a bit like songs, rhymes or magic spells. It certainly goes back to growing up in a smallholding in North Ayrshire called Wheatiefaulds, surrounded by places with funny sounding names, like the Bumbo burn (Bombo on the map), Ashiefalls or 'the Crunchy' ferm (Kerselochmuir), the 'Tiddles' (Todhill) or Tottery and Tillietudlem Castles, names our Dad, Willie Hodgart, used for nearby ruins (Tillietudlem, a village near Carluke, is the fictitious name for a castle in Sir Walter Scott's novel *Old Mortality*, supposedly based on Craignethan Castle, near Lanark).

He told us that these 'castles' had been 'focht owre mony times' by the men of a former mining village, the Den, which once had a fearsome reputation. I was really quite disappointed when I learned years later that they were only auld pit ruins or bourachs. He was at least telling a poetic truth about the auld ruins having been 'focht owre', but at least the nearby bings were built by the Den men, though it certainly wasnae on their holidays so that they could see the sea, anither o his tall tales!

So my interest really started from where I grew up and learning not just that places roon aboot sounded funny, which obviously made them stick in my mind, but that these places held memories of people and had a history behind them. But of course I discovered much later that their names and their histories were sometimes deceptive or unreliable for one reason or another. Yet surely every place has its stories, myths, legends and folk etymologies if we dig deep enough, an that is the pynt o ma wee story aboot whaur I grew up, for I think we should be helping oor weans tae seek oot an dig up stories like that, something that is a key element in the classroom activities on our website.

Someone I'm sure who would have agreed with this idea was our great novelist Neil Gunn whose writing resonates with a strong sense of place and its powerful influence on the lives of Highland folk, as shown in his novel *Young Art and Old Hector* (Souvenir Press, 1976), where the old man tells the young boy: 'I know every corner of this land [...] and I know of things that happened here on our land long, long ago [...] Every little place, every hillock, every hill and slope, has its own name' (p. 250).

Discovering the meaning of a place has been compared to unlocking a moment in time, or rather a place in time, like finding an insect trapped in amber, a very apt metaphor indeed, but perhaps it can also be seen as something more alive if we visualise or imagine the landscape, as Sorley MacLean does in his great poem 'Hallaig', as a living entity, a web of interconnectivity, alive with presences and perceptions from past times.

I would also suggest that an understanding of place-names can maybe allow us to tune into the land in a deeper way, as it can let us hear the words, the voices and even the mindset of these long ago folk, something that may well be the beginning of a lifelong engagement or dialogue with our landscape and the people who once lived there, for, as Nan Shepherd has shown in her remarkable book *The Living Mountain*, the hills are indeed alive! It is also in many ways a form of reclaiming our land from the mapmakers who often excluded not only the voices of the folk who lived in places they were mapping but were also deaf to the voices of those who gave names to places long before maps were ever made.

The realisation that all place-names mean something, or have a story behind them, is actually quite an important step in becoming more aware of our environment, history and languages and an even more important step is eventually realising that names often don't mean what they seem to say, or perhaps seem to say what they don't mean! That is because maps can perpetuate myths and mistakes, or at best half-truths, as the men who made the maps usually didnae speak the local lingo! And we also have to reckon with folk etymologies, which should not always be discounted, as they are part of the story too. However, things are often not what they seem, even when the name might apparently seem as obvious as 'Bearsden'. Various meanings have been in circulation over the years for this town (originally called Kirktoun) in East Dunbartonshire, but it certainly does not mean the den of the bears.

The most common explanation is that the sons of a local laird once kept a pet bear in a den there or that it's a nickname given to the area of the Manse Burn by a Garscube heir, but there is no evidence to support either explanation. More credible possibilities are that it could be from Gaelic or Brittonic, meaning 'entrenchments of the fort', or it relates to the barley (*bere*) that grew in the glen (*dene*), which is possible, but it is more likely to come from the Old English word *bar*, meaning boar and *denu / den* which means dell or valley, so it probably means the dell or den of the boar.

Yet there is nothing new about arguments over place-names and some probably go back a very long way indeed. The function of place-names may originally have been just to locate a place and find the way, a bit like landscape markers or signposts, then later to honour someone or something sacred, or simply describe what it looked like, what happened there or what it was used for, and later to show who had power or control over it, a bit like animals marking out their territory or gangs painting their graffiti on walls! Tribal oral traditions preserved names for centuries, or even millennia, long before chroniclers or mapmakers arrived on the scene, though names must have changed a lot as they were passed on from generation to generation and the language, culture or the environment changed, something that no doubt caused disputes over what some places should be called.

The renaming or misnaming of place-names has of course happened the world over as place-names were translated from one language to another, the original language lost or deemed uncivilised and unintelligible. This was often because colonisers didn't just occupy, but renamed the landscape as part of a political process of eradicating the local tongue and culture, usually renaming places with something they considered more 'civilised' or more prestigious, like the name of their discoverer, their general or monarch, as happened throughout the Roman or British Empires and many others.

In our case, place-name corruption probably started even before the Romans arrived and these visitors certainly left their mark, as Greek and Roman explorers, historians or cartographers put names to places at the end of their known world. Our main source of early names is from the fourth-century BC Greek navigator Pytheas, who, for example, gave the name *Orcas* to the Orkney Isles and *Hebudes* to the Hebrides, while in the first century BC the Greek historian

Diodorus Siculus refers to an island called *Hyperborea* ('beyond the north wind') and the earliest Roman geographer, Pomponius Mela, around AD 43, is said to be the first to name, and locate correctly, the Orcades, or Orkney Islands.

Roman military campaigns gave us quite a few of our names, as recorded by the historian Tacitus in the late first century AD, such as *Mons Graupius* (possibly a misprint for Grampius) or *Clota* for the Clyde (their version of a name the locals used), while Ptolemy's maps, from his great Geographia of around AD 150, give about sixty Scottish place-names. Around two thousand years later, some of these names are still used in various forms and undoubtedly those Mediterranean visitors, especially the Romans, have a lot to answer for!

Other early sources are mainly ecclesiastical documents from the seventh and eighth centuries AD onwards, especially monastic charters or the famous chronicles by St Bede, a Northumbrian monk and scholar known as the 'Venerable Bede'. Various Brittonic / Brythonic names are found in the *Book of Taliesin* (a medieval manuscript based on fragments and legends of a sixth-century Brittonic bard or bards) or in *The Gododdin*, an epic poem of around the same period (attributed to a bard called Aneirin or Aneurin), while Gaelic names appear in *The Pictish Chronicle* (possibly tenth century) or *The Book of Deer* from the eleventh and twelfth century, or later in John Fordun's Chronicle in the fourteenth century.

Many names are found in abbey records and legal documents from the early Middle Ages specifying land holdings and boundaries held by the new Norman feudal nobility, e.g. the English king Edward I's Ragman's Roll of 1296, while we also have burgh records and charters, such as the foundation Charter of the Royal Burgh of Ayr in 1205 which is full of Gaelic place-names, as is the rest of the county, something that might surprise many folk.

We also find many references to place-names in John Barbour's fourteenth-century poem *The Bruce* and Blind Harry's fifteenth-century poem *The Wallace* in Scots, or in the works of writers and scholars, like Dunbar and Henryson, or Hector Boece's *Scotorum Historia* (1526) or George Buchanan's *Rerum Scoticarum Historia* (1582), or the Rev. James Fraser's Wardlaw Manuscript of about 1674, but none of these are primarily interested in place-names. The first attempt to survey Scotland properly was made by Timothy Pont between 1596 and 1614, but unfortunately his mapping was incomplete: some of the maps were lost and they were not properly published until the nineteenth century.

Yet Pont's maps, revised by Robert Gordon of Straloch, formed the basis of the first known Atlas of Scotland, but made by a Dutchman, Joan Blaeu, written in Latin and published in 1654, which is a goldmine of Scottish place-names. Most of the Gaelic names were given phonetically in Scots by Pont (i.e. as sounded in Scots, but not spelled in Gaelic) with reasonable accuracy, so he must have understood Gaelic or had Gaelic translators. In the eighteenth century, General William Roy and others made accurate maps but they were mainly interested in places or features that had a military significance, while most of the islands were omitted and Gaelic names were seriously distorted. Ordnance Survey maps, begun in Scotland in 1819 though not complete till the end of that century, were more thorough, but they also sometimes got the names wrong or perpetuated corrupted versions.

While many Gaelic names in the Lowlands have been Scotticised, i.e. changed into something that made more sense to speakers of Scots, many Scots names were in turn anglified or replaced with English ones from the eighteenth century onwards, as the upper classes, believing that Scots was 'incorrect' or 'bad' English, set out to 'improve' their language by removing 'Scotticisms', something commented on by an Aberdeen minister in the 1790s:

> We cannot give a better example [...] of the advances [...] which we are daily making towards English. We almost never hear now of the Braidgate and the Castle Gate. They are become universally the Broadstreet and the Castle Street [...]
>
> —John Sinclair, *The Statistical Account of Scotland*, 21 vols (William Creech, 1791–99), XIV, p. 298.

Likewise *gates* and *wynds* were often changed to *streets* and *lanes*, as in St Andrews where a nineteenth-century provost was determined to raise the status of the town by anglifying Scots street names, e.g. changing Baxter Wynd to Baker Lane, Mercatgait to Market Street and East Burn Wynd to Abbey Street, which no doubt met with the approval of the town's gentry, as did changing the name Glabertoun (meaning mucky farm) near Edinburgh to Clermiston, in the eighteenth century.

However, most of the misnaming or renaming of our place-names has been via translation from Gaelic, firstly into Latin, then into English or Scots, as incomers and mapmakers corrupted words they didn't understand. In addition, place-names often go through a process of folk etymology, or familiarisation, whereby unfamiliar or difficult names are gradually altered to something more familiar or pronounceable, e.g. the Brittonic name *baedd* or *bod coed* (boar wood) eventually transformed into Bathgate. Yet sometimes, in spite of the spelling, the local tongue has preserved the original name, or something close to it, for a very long time indeed, as in Glesca, Pertick, Mullgye or Fawkirk.

Bearing all this in mind, it is hardly surprising, therefore, that nearly every officially sanctioned new place-name in mainland Scotland over the last few centuries has been in English, not Scots or Gaelic. I cannot think of another European country which has allowed its indigenous languages to be so inconspicuous in or missing from its public place-naming.

2. WHY STUDY PLACE-NAMES?

> Place names haud the history o a people. Gin we dinna ken the words naming the land aboot us an the places whaur we bide, we dinna ken Scotland. Therefore, the current weirin awa o place names maun be coontert as a maitter o urgency.
>
> —'Scots, A Statement o Principles', published by the Scots Pairlament Cross Pairty Group on the Scots Language, 2003

Like the members of the above group, I believe a basic knowledge of place-names is a key component in understanding our environment. Learning should surely start from where we are and work outwards in ever widening circles emanating from the centre, and a key part of that centre is where we live, the place we call mo dhachaigh / hame / home, the people and places around us.

For schoolchildren, the study of place-names is an ideal topic for interdisciplinary learning, as it runs across so many subjects and of course the Scottish Studies Certificate depends on a cross-curricular approach. The Scottish Studies Working Group (2012) certainly felt that it was an ideal core topic, as it can bring together so many areas of knowledge to develop a more integrated understanding of not only where we live but also hopefully develop a sense of where we have come from, who was here before us and how they left their mark: a geographical and historical exploration of a rich environmental and linguistic resource on our very doorstep.

As well as learning about our rich linguistic history, learning about place-names will also enhance our children's language skills, including the skills for learning other languages. In addition, it should also allow our children to discover that our native languages, including the local vernacular, have a linguistic heritage and legitimacy that has often been denied by our educational system in the past, a discovery that should help young folk growing up in Scotland to appreciate that the local speech is not 'bad' English or 'incorrect' pronunciation. In fact it often preserves the 'correct' name for the place and it is actually the modern pronunciation or spelling which is wrong, mainly due to a process of cultural displacement which involved demeaning the native speech of a place and branding it as 'barbarous' or 'ignorant'.

Learning about place-names can also be seen as an act of reclaiming both the landscape and the language our place-names are written in, or should have been written in, giving legitimacy to the voices of folk who have too often been ignored or disregarded. Studying our place-names should also help all of us, but especially teachers and students, to appreciate that Scotland has long been a multilingual and multicultural wee country. In other words, we were a multi-ethnic polyglot lot from the time of MacAdam and we have long benefited from incomers settling here and making Scotland their home.

3. A GUIDE FOR TEACHERS AND STUDENTS

The study of place-names and their meanings is known as toponymy (from Greek *topos*, a place, and *onoma*, a name), while the term hydronymy is sometimes used for water or coastal features (from Greek *hydor* = water). A related word is onomastics, which means the study of the origins, history and use of proper names, and toponomy is one of its main branches.

Place-names have been described by a famous scholar of the subject, Professor Bill Nicolaisen, as 'linguistic archaeology', often involving a process of sifting through or excavating earlier forms or layers, and by studying the language, or languages, which certain names come from, we can learn quite a number of interesting things:

a) the age of the name
b) the people who gave it that name
c) what it originally meant
d) perhaps what once happened here
e) perhaps how it has been changed throughout our history.

Scotland's place-names often look quite mysterious or sound quite funny, with wonderful sounding names like Blinkbonny, Clachnacuddin or Drumnadrochit, names derived from our rich and diverse linguistic history as well as from the distortion or corruption of these names. Many of our names come from Gaelic, which was once spoken over most of the country, and when we look closely we will find that many of these names often give us very clear, accurate descriptions of the place and its topography (natural features, description of the landscape, such as hills, rivers or woods).

These original names usually meant something very exact, giving accurate information to guide the traveller or to remember them by, with vivid descriptions of places or features, often using comparisons to the human body, people or animals, e.g. *bodach*, an old man or spectre, is quite common, and sometimes found in pairs, as in Tigh nam Bodach (house of the old men) and Tigh nan Cailleach (house of the old women) in Glen Lyon, or Ceum na Caillich, the witch's, or old hag's step on Arran, and there are just as many old women as old men in the landscape! In fact, when you are investigating the meaning of a name you should, if possible, always try to check it against the local topography to see if it makes any sense or if there is any landscape evidence for the name from hills, rivers, coastlines. This could easily tie in with other environmental or outdoor activities and projects.

Clear maps of the local area or the whole country are essential, while Gaelic and Scots dictionaries and handy glossaries of place-names would be very helpful, many of which are accessible online.

The Spelling is Atroshus!

It is important to stress in studying place-names that the original language and its meaning have often been changed, sometimes beyond recognition, mainly by English-speaking mapmakers, and so the spelling often bears little resemblance to the original, such as Kingussie (pron. Kinyoosie) which is from the Gaelic *Ceann a' Ghiùthsaich*, meaning head of the fir wood, while Applecross has nothing whatsoever to do with orchards, but is a corruption of *aber crosan*, meaning mouth of the River Crossan.

The Isle of Arran has some very funny names indeed, such as Thundergay – which might suggest a place where folk were once very happy during thunderstorms. But, of course, like all our islands, Arran can be a very windy place and the name comes from the Gaelic *tòrr na gaoith*, meaning hill of the wind or even *tòn ri gaoith*, possibly the same as Tundergarth in Dumfries and Galloway, backside to / of the wind! And Shedog, on the same island, isn't a place where you might have gone to buy a female dog, but just another blowy place, from the Gaelic word *sèideadh*, blowing, or from *sèideag*, a little puff!

Yet another windy place is Curly Wee in Dumfries and Galloway, which does not describe what the natives of the area once looked like but refers to the hill being a windy corner or bend, from the Gaelic *cuir le gaoith* (literally meaning bend with the wind), while a damp place never short of wind is the Bog of Gight in Aberdeenshire (from *gaoithe*, of wind). In the same shire there is even a place called Brokenwind, which might be from Gaelic *broc* or Scots *brock*, a badger and a wynd, though that is maybe a wind-up! At least Guthrie, near Arbroath, is just a windy slope, from Gaelic *gaoth* and *ruighe*, but there is no doubt that Windygates and Windy Yett are still windy roads and Guay in Perth is just full of wind!

Many people might have reservations about living in a place called Sodom on the island of Whalsay, Shetland (though it didn't deter the poet Hugh MacDiarmid from living there for a time), but it does not however mean that it was once seen as a very wicked place or tell us what their neighbours thought of them, as the name is from an Old Norse or Shetlandic (Norn) word, *sudheim*, simply meaning southern home, much to the relief of visitors.

Yet at least some people in Orkney and Shetland might not be too offended if you called them Twatts, as that is actually where they come from (Old Norse *thveit*, a clearing or meadow), and you would only be doing the same if you called folk in an Aberdenshire village Clatts (from Old Norse *klettr*, rocks, cliffs), but Twechars would not be at all offended, as they proudly belong to this former mining village, near the

Antonine wall, possibly via a series of sound shifts from the Gaelic *tuineach*, a dwelling.

You might also think it was a very bad idea to visit various places called Badcall or Baudcall in the Highlands but they are harmless enough places as they are only patches or thickets of hazels, from Gaelic *bad* and *coll*. Admittedly we do have a few crooked places, but neither Crook of Devon nor Crook of Alves have criminal records, as they simply come from the Old Norse *krokr*, a bend or crook on a river.

Thankfully, you don't have to be really tough to live in Kirkton of Tough in Aberdeenshire as it is only a hillock, from the Gaelic *tulach*, and Tuskerbuster in Orkney is not a local tough guy, but only a peat cutter's farm, from the Norse *torf*, *skeri* and *bolstadr*. However, people in Brawl or Braal on the north coast will certainly assure you that they are not always fighting on their broad hillside (possibly from Norse *breithr hallr*) and Yell folk in Shetland are not as noisy as they seem, as they live on a barren island, from Norse *geldr*, probably of barren or calfless cows, though the farmers would not have been happy about that.

You would also be very mistaken for thinking folk in Spittal or Spittal of Glenshee have some very unpleasant habits, when in fact they offer refuge or shelter (from old Gaelic *spideal*), and Dinnet in Aberdeenshire is not a place where people have left very few things undone, but is also another place of refuge, from the Gaelic *dion*, shelter or defence, plus *àit*, a place. Happily, Scotstarvit in Fife was not named after a hungry or badly nourished Scot but a farm where a man called Scot, or Scots people, once had a bull, from the Gaelic word *tarbh* (pron. *tahrav*).

Maybe Dullatur, near Cumbernauld, and Dull, in Aberdeenshire, are not as boring as they sound (in spite of the fact that Dull is twinned with Boring in the USA), though the former is indeed a dark slope, from Gaelic *dubh* and *leitir*, and the latter is only a field or meadow (Gaelic *dail*). Other places that don't sound too clever are Dowally (but not doolally), a dark cliff, from *dubh* and *aille*, Dunino, a moor hill, from Gaelic *dùn* and *aonach*, and Dumyat, from *dùn* or hill of the Maeatae, the name of the local tribe according to the Romans.

Certainly Macmerry in East Lothian sounds a lot happier, even though it may be a plain place (from Gaelic *magh*, meadow or plain and possibly the old Gaelic *meurag*, a small pebble, or finger, or maybe *Moire*, Mary's), but Bonkle in North Lanarkshire and Buncle in East Lothian could well be a comic double act, and they are at least related, as they both maybe come from the Gaelic *bonn chille*, bottom church. Unfortunately Boddam in Aberdeenshire, Boddin, near Montrose, and even Boon in the Borders will never rise to the top as they are all bottom places (Old English *botm* and Gaelic *bonn*), though their residents will no doubt argue otherwise. At least their spelling isn't too bad!

On the other hand, many people might be delighted to live in a place called Blinkbonny (often used for house names), though it is just a milk farm or village, from the Gaelic *baile bainne*. Yet there could be a lot of blinking going on in Bonnybridge as apparently a lot of UFO sightings have been reported here and hopefully it is a place built to last, as it could be from Gaelic *buan*, meaning lasting, enduring, though maybe it was a place where you once had to be very careful crossing the river, as the burn was very fast-flowing, from Gaelic *buain*, to cut, prune, pluck or pull; thus a river that would soon sweep you away, though maybe today that is more likely to happen with a flying saucer! The river name passed from the bridge and then to the town, a name which has certainly lasted.

Yet the bishop of Bishopbriggs maybe didn't mind too much not having a bridge as he was quite happy with his *riggs* or fields, while Carstairs isn't a place with an escalator for vehicles, but the castle of a man named Tarras, possibly a bishop, so he maybe did have a fancy carriage or cart at least! However, Minigaff in Galloway is not a wee mistake as it was once the hill of the smith, a name derived in a very roundabout way from the Gaelic *monadh a' ghobhainn*.

Various formidable-sounding women have long inhabited our land, but Maggieknockater in Moray doesn't refer to a woman called Maggie who was renowned for knocking on people's doors, maybe a sort of chapper-up for getting folk to work on time, but derives from the Gaelic *magh*, meaning a plain, and *an fhùcadair* (pron. ahnookiter), meaning 'of the fuller', someone who cleans and thickens cloth. However, Megginch was not a tiny woman called Meg but a milk meadow or island, from old Gaelic *melg*, milk, and *innis*, island, meadow. Sadly there was no lady in Ladybank, though hopefully there are a few now, but it once was a soggy slope, from the Gaelic *leathad* and *bog*!

We also have several Mauds up in the north east, including an old one, but they have all gone to the dogs as they are from the Gaelic *madadh*, a dog, or possibly meeting places, from *mòd*, though maybe they were places where dogs met! Likewise a part of Glasgow was not named after a legendary or infamous woman called Bella Houston, but named after a very holy spot indeed, the village of the crucifix, from the Gaelic *baile cheusadain*, while Belladrum was not a woman who liked beating her drum whenever things got up her hump, because it was only a humpy place at the mouth of the ford, from the Gaelic *beul àtha druim*. At least Maryhill was a real Mary Hill who was in fact the landowner.

Some may well have been rather disappointed to find that Bellochantuy in Kintyre was not an Italian or French woman who was a beautiful singer but just a place where there was a house at the pass or gap, *bealach an taighe*. Yet Girlsta in Shetland is indeed where a girl (one Geirhildr, a Viking explorer's daughter) stayed, as this was the girl's dwelling or homestead, from the Nosre *stadr*, though Manish in Harris isn't a place where even the woman are very macho, as it is in fact a gull headland, or ness, from Old Norse *mar* and *nes*.

However, there is no shortage of men lurking in the landscape, such as Manuel in West Lothian, not named after a Spanish waiter, but a rock view, from Cumbric *maen* and *gwel*, while no Italians lived in Turin, near Forfar, when they called it that, as it was only a *torran*, a wee hill. And neither were there any Turks in Brig o Turk, as it was possibly only a pig's bridge, from *tuirc*, of a pig or boar. What a bore indeed to have a bridge named after it!

Flashader in Skye might seem to be a place for flash guys or show-offs, but it's only a flat farm, and, although Blackadder in the Borders sounds a bit venomous, you won't find any black snakes there, or even a comic character portrayed by Rowan Atkinson. Like its neighbour, the Whiteadder, you will only find water there, as it was given its name by Celts in days of yore, possibly referring to a water god or spirit, with probably the same name as the Oder in Germany, so possibly a sprite that has been lurking around rivers for millennia. Things are indeed not always what they seem or seem to say, something that has helped to give rise to folk etymologies or popular explanations of a name people didn't understand, many of which are memorable or funny, but rarely accurate.

But Listen to the Locals

Local pronunciation, including stress patterns, has often survived centuries in spite of English spelling, or rather misspelling, and is in fact very likely to provide a better guide to how the name should be said and what it originally meant than what it says on the map, e.g. Milngavie is still pronounced Mullguy, as it is probably a contraction of the Gaelic *muileann*, meaning a mill, and either *gaoithe*, meaning of wind (pron.

geuee) or *Dhai* (pron. guy), a shortened form of Davy's, so it could have been a windmill or Davy's mill, or alternatively, it could be *meall na gaoithe*, meaning bare hill of the wind. And the nearby town of Drymen, sometimes mistaken by visitors as a place where the men are often thirsty or never go out in the rain, is called 'Drimin' by the locals as it is really a place on a ridge (from Gaelic *druimean*, or *druman / droman*, a wee hill or ridge).

Likewise, the name 'Glesca', which children were once reprimanded for using in school, is possibly more 'correct' than Glasgow, as the name probably comes from Brittonic *glas*, meaning stream or green (for landscape), and *cau*, a hollow, though other meanings have evolved, as *glas* can also mean grey (for people and animals) and *cù* can mean either dear or a dog, though maybe it was a dear dog or named after somebody's greyhound, possibly St Kentigern's! It seems to have been Jocelyn of Furness, who wrote a *Life* of St Kentigern in the late twelfth or early thirteenth century, who added the *cu* suffix to reinterpret the meaning as a dear green place / hollow. Similarly, 'Pertick' is much closer to its Brittonic origin, *perth*, a thicket or bushy place, than Partick, while Falkirk is indeed 'Fawkirk', as it comes from the Scots *faw*, meaning speckled, which is a translation of the earlier Latin (*varia capella*) which in turn was taken from the older Gaelic, *eaglais bhreac*, meaning speckled church, as it was built of variegated stone.

Another memorable example of this is the Fife name Kilconquhar which is actually pronounced 'Kinyuchar'! This is because it was originally Cill Conchubar (an old Irish name, especially of Irish kings, meaning a dog lover, the origin of Connar or O'Connor names), which became shortened to Conchar and eventually to *euchar* and the *cill / kil* was dropped but the *con* changed to *kin*, so 'Kinyuchar' is definitely closer to Conchubar than Kilconquhar! Likewise locals don't pronounce Deskford, in Aberdeenshire, as it is written, but call it 'deskart' which is much closer to the original Gaelic *deas ghart* (south enclosure or field).

No wonder place-name scholars like to find the earliest written versions of names in legal documents, church charters or literary texts in their efforts to trace or verify the original form of a name, though even these are not always reliable as they very much depend on who was writing them, how they spelled them and whether they understood the language of the original.

Thus, tracing the original name can sometimes be very puzzling, especially if there is no historical or written evidence and if there are no landscape or environmental clues.

4. THE STRUCTURE OF PLACE-NAMES – THE BUILDING-BLOCKS

Place-names usually contain two parts or elements: a **generic** or **classifying element** and a **qualifying** or **specific element** or **elements**.

The **generic element** gives us the type of place or general category it can be fitted into, e.g. house, church, hill, river. This can sometimes be made up of more than one word, thereby forming a new **compound** word (two elements put together) e.g. *milntoun*, meaning mill farm or mill settlement, often contracted to 'milton', made up of Scots *miln* and *toun*.

The **qualifying** or **specific element** gives us additional information to describe or explain, often with reference to colour, shape, flora and fauna, location, function, building or person, e.g. Kilwinning, which means the church of St Finnian. The generic element kil is from the Gaelic *cill(e)*, meaning church or cell / shelter lived in by a monk. The qualifying element is the name of the monk who lived there, or saint to whom the church was dedicated, as in this case, Finian or Finnian, an early Celtic saint.

The lists in this guide concentrate on common **generic elements** because if we can identify these, we can work out which language they came from and hopefully learn quite a lot about them, even if we sometimes can't work out their full or original meaning.

In names which are of **Celtic origin,** the generic element normally appears first in a compound name, as in Dumbarton or Kilmarnock, though this is not always the case as some examples place the specific / describing element first for emphasis, e.g. Garbh-Bheinn, a *rough* mountain, while older names often do likewise, maybe for the same reason, e.g. the Dusk Burn in North Ayrshire or Duisk Burn in South Ayrshire (both dark waters, from Gaelic *dubh-uisge*, or from the earlier Brittonic equivalents, *dwr* and *usc*) or Duart in Argyll (a dark height or point, from *dubh-àird*).

However, in names of **Germanic origin** (Norse, Anglic and Scots), the specific element comes first and the generic element comes second, as in Orkney, Lerwick, Goat Fell, Borrodale, Bannockburn, Hamilton, Houston, Hampden, Philiphaugh, Greenloaning.

Well over half of our place-names are Celtic (Brittonic / Brythonic and Gaelic), though much changed by Scots or English spelling, but other languages have also left their mark, sometimes overlapping each other and often leading to scholarly disputes over the origins of the name. A high proportion of Scottish place-names are prefixed by Ach, Bal, Dun, Ken and Kil and many have the suffix ach / och, dal / dale, ton / ington. If all such words had only one meaning, they would be a lot easier to make sense of, but unfortunately they sometimes have several possible meanings, thereby making things a lot more complicated.

Names have also often undergone a process of transmutation or combination to form complex hybrids or even new life forms, as in the Kilconquhar example above, while we often have to look out for examples of **metathesis** (the transposition of letters, sounds or syllables within a word), as in *car* / *caer* to *cra* in Cramond, *tir* to *try* in Tillicoultry, or the replacement of a word with another similar-sounding one, as in *caer* to *kir* or *kirk*, *ceann* to *ken* or *kil*, or *cùl* and *caol* to *kil*, or *mòr* to *muir*, *cardden* to *càirn*. Names also sometimes suffer from linguistic erosion, compression or elision with the passing of time, so that words are shortened and letters missed out, such as Anstruther to Ainster / Enster in Fife, or Bethocrule to Bedrule in the Borders.

In many ways, a name often captures a moment or period in the history of a place and we can sometimes see this in place-names appearing and disappearing, forgotten or changed. Indeed, an English landscape scholar, W. G. Hoskins, has compared place-naming to the study of palimpsests, i.e. documents or paintings which have been overwritten or overpainted many times, another very apt metaphor indeed. However, maybe we could also see place-names study as a bit like trying to open a very ancient and mysterious book, but needing the magic words to open it via an understanding of the languages it is written in.

Our place-names actually come from eight or nine languages: Pre-Celtic, Pictish, Brittonic / Brythonic / Cumbric, Gaelic, Norse, Latin and French, Anglic and Scots. However, they can be conveniently grouped into about five main categories: Pictish, Brittonic, Gaelic, Norse, Anglic / Scots.

For the most part the names discussed in each section will be divided into those describing **physical** or **landscape features** and those describing **settlements.** Settlement names are particularly important because they are clear proof, firstly, that this shows which language had status or power at the time of naming, and, secondly, that the speakers of the language actually lived in the area for some time and weren't merely visitors or tourists! Settlements were often very small and the word 'village' could simply mean a small group of families.

Other Categories or Classifications

Place-names can also be categorised according to activities or purposes, such as:

> **Commemorative Names** (for somebody or something important): Fingal's Cave, Waterloo, Ladysmith Road, Charlotte Square, George V Bridge.
> **Ecclesiastical or Church Names:** many Kil and Kirk names, Canonmills, Chapelton, Holyrood, Monkton, Prestwick, Terregles.

- **Evaluative or Reflective Names** (suggesting an assessment or judgement, possibly sarcastic): Cauldhame, Morningside, Sunnybank, Mount Pleasant.
- **Fiscal / Tax Names:** Bawbee Brig, Merksworth, Merkland (bawbees and merks were old Scots coins), Pennyland, Penninghame (penny holding or rent), Poundland, Fortyshillingland, Oxgangs (as much as an ox could go, or plough in a day).
- **Man-Made Structures:** Duntocher, Greendykes, Newbigging, Newbridge, Newmilns.
- **Occupational / Work Names:** Prestonpans, Saltcoats (both from the salt-making industry), Dockheid / head, Wauk Mill or Fullerton (both from waulking or fulling of cloth), Fisherton.
- **Personal Names:** Balmalcolm, Dundonald, Gilmerton, Johnstone, Winchburgh.

There are several variations on this, such as the seven main types referred to in David Ross's *Scottish Placenames* (pp. xvii–xviii), a very useful short dictionary of Scottish place-names with an excellent introduction (see **Bibliography**).

A more recent categorisation, or way of viewing and understanding the landscape, is that devised by a Dutch scholar, Meto Vroom, which is outlined in John Murray's fascinating book on *Reading the Gaelic Landscape / Leughadh Aghaidh na Tire*: to think of the landscape in terms of three overlapping and interacting horizontal levels or layers: the abiotic or non-living at the bottom, i.e. the physical landscape (rocks, mountains), the biotic or biological (plants, water, animals) in the middle and the occupational or cultural level at the top, i.e. humans and their way of life.

A model like this takes us into environmental, biological, botanical, geological, historical, geographical, ethnic and cultural studies and various other related subjects or disciplines, as it attempts to view living systems as a whole to understand how all life is part of a complex interrelated ecosystem with one aspect impacting and affecting all the others in a great web of life. This is something Nan Shepherd fully appreciated in *The Living Mountain*. Change or damage to one part affects all the other levels, such as changes in land use with the burning and clearing of woods, replacing mixed farming communities with large sheep farms or sporting estates during the Highland Clearances, or covering the landscape with huge blocks of pine forests, thereby destroying the biodiversity of the environment. Changes like this also have a great impact on how the landscape is seen and how it is named.

Therefore as places and people change over time, names are often made up of several elements, resulting in compound place-names which often have more than one of the above functions or combine two or more languages. Thus we have many **hybrid** names which are partly from one language, partly another, as the speech of incomers was grafted onto older words and names, as with older Brittonic or Gaelic names with a later Scots or English part added, e.g. Bannockburn or Scotstarvit, where Scots has borrowed or incorporated older Gaelic words, though there are even a few examples of a Gaelic element being added to an English name, possibly as in Tarbolton in Ayrshire (See OE **botl** / **bothl**). We find the use of these loan words across the country, but they are especially common in the Lowlands.

Names like the above often help to uncover a particular form of history which geographers call 'sequent occupance' i.e. the occupation of a landscape by a series of peoples with different cultures, languages and maybe different ways of using the land, which leave some kind of legacy in the present landscape and so need to be taken account of to get a full picture of what produced the landscape we see today.

Thus we sometimes find, to use the jargon, **pleonastic** names (superfluous or unnecessary words with the same meaning, i.e. place-name tautologies) where incomers grafted their word for a place onto an existing name, which meant exactly the same thing, because they did not know what the existing name meant. Thus we get tautologies like Eas Fors, a waterfall on Mull, where we have both the Gaelic word *eas* and the Norse *forss* for a waterfall, while on the same island we find Quinish Point, from the Norse *kvi*, a cattle pen, plus *nes*, a headland or point, to which has been added the needless point! We also have Knockhill, a hill hill, Ailsa Craig, possibly a rock rock, Ardnish Point, a high point point, Castlecary, fort of the forts, similar to Dunreay, hill fort of the circular fort, or Fetterletter, a slope slope, both parts being Gaelic, all of them saying the same thing twice.

Therefore, bearing all of the above in mind, we frequently find our place-names much disputed because, like places called Threipland, we often live on debateable ground, so cautionary words are advisable, like 'possibly', 'perhaps' or 'could be', or, with more certainty, 'probably', 'more likely' or even 'mibbies ay, mibbies naw'! As they say in Gaelic, it is often a case of 'cò aige tha fios?', or in Scots, 'wha kens?'

5. INDEX OF KEY NAMES AND THEIR LINGUISTIC ORIGINS

This is an index of the words listed alphabetically in each language section to follow. The original or underlying word is given first, followed by any reflexes or variations derived from it, plus any similar words (in meaning or just in sound) which are found in the same section.

Grave accents are used in some Gaelic words to show that the vowel is a long one. The following abbreviations give the section(s) a word is listed under: P = Pictish, B = Brittonic / Cumbric, ON = Old Norse, OE = Old English, S = Scots.

A
Aa, aar, air & ayre / ire (ON)
Aber (P & B)
Abhainn, abh (G)
Achadh, auch, ach (G)
Àirigh, àiridh (G)
Al / ail (B / G)
Allt (G)
Annaid, annat (G)
Aonach, aoineadh (G)
Àrd, àird (G)
Àth & fadhail (G)

B
Bac (G) – see clais
Bad, badan (G)
Bàgh, bàigh (G)
Baile, bal (G)
Bakki (ON) – see brae

Bàrr (G)
Bealach (G)
Beinn, ben (G)
Bekkr, bek, back (ON)
Bent(s) (OE / S)
Biggin(g) (OE / S)
Blaen, blane (B)
Blàr, blàir (G)
Bog & riasg & fèith (G)
Bolstadr, bo, boll, bost (ON)
Bonn & bun (G)
Borg, burra (ON) & burg, burh, burgh (OE / S)
Both(an), bùth, bùthan (G)
Botl, bothl (OE) as in bol (ON)
Brae (ON), bràigh & bruach (G), bakki (ON)
Brig(g) (OE / S)
Brothag (G) – see clais
Bru / bru'r (ON)
Bur (ON / OE)
Burn (ON & OE / S)

C
Caer, cair, cader, cathair, kir (B / G) – see also ràth
Camas, camus & chaim, crom & lùb (G)
Caol, caolas, kyle & cùil & cumhang (G)
Cardden (P / B)
Càrn, cairn, cùirn & càrnach (G)
Carreg, carraig (B / G) – see ail & creag
Cau, cow & coomb / cwm (B)
Ceann, ken, kem, kin & ceap (G)
Ceap, ceapach (G)
Cefn, cyffin (B)
Cill(e), kil (G)
Clach, cloiche & leac (G)

Clais, cos, coise & bac, lag, glac, brothag, toll (G)
Claon (G) – see leitir
Clett / cleat (ON) – see hus
Cleugh, cleuch (OE / S)
Cnoc, knock (G)
Coed / coet (B)
Coille & craobh & doire (G)
Comar (G) – see inbhir
Corran (G) – see sròn
Corrie / coire (G)
Cot (OE / S)
Creag, creg, craig, carraig, ail (G)
Creek, gja / geo / goe (ON) – see fjord, firth
Crìoch (G)
Cuach (G)
Cuan (G) – see traigh
Cùil(e), cùl (G)

D
Dail, dal (G), dol (B) & lèana, lèanag & àilean (G)
Dale (OE), Dalr (ON)
Davoc, davoch, doch (G)
Den, dean (OE / S)
Dobhar, dwr (B)
Dod (OE / S)
Doire (G) – see coille & also bad
Dòirlinn (G) – see tràigh
Drum, druim, dron (G)
Dùn, din, dum, dunadh, dìong (B & G)

E
Earrann, arn / iron & fearann (G) – see also roinn
Eas (G) – see uisg(e)
Edge, ege (OE / S)

Egles (B)
Eilean (G)
Erne (OE)
Ey, ae, ay, eidh & aith (ON)

F
Faithir (G) – see leitir
Fas, fasadh & fasgadh & fàsach (G)
Fauld, fold (S / OE)
Fell, fjall (ON)
Fjord, firth & gja, geo, creek, krokr (ON)
Fuaran (G) – see tobar

G
Gafael (B)
Gardr / garðr / garth / girth (ON) – see also gàrradh
Gàrradh & geàrraidh & gairt (G) – see also gardr
Gil / ghyll (ON)
Gleann, glen & glac, slochd, sloc (G)

H
Haa, haw (OE / S)
Hag(g) / hags (OE / S)
Haining (ON)
Ham (OE / S)
Hamn, hofn (ON) – see hop & vik
Hang & hofud, aith (ON) – see ness
Hauch, haugh (OE / S)
Heuch, heugh (OE / S)
Holm, holmr (ON & OE / S)
Hop, hope (ON) & òb & hamn, hofn – see also vik
Howe (OE / S) – see hop
Hus, housa, hoose (ON & OE / S)

I
Inbhir, inver & comar, ùidh (G)
Innis, inch (G)

K
Kame(s), kaim (ON & OE / S) – see camas
Ken, kin, as in ceann (G)
Kerse, carse (OE / S)
Kil as in cille (G)
Kirk, circe (ON & OE / S)
Klettr, klakk, knappr, kollr & hella / hellya (ON)
Knowe (OE / S)
Krokr (ON) – see fjord
Kyr, kye & kvi / quoy (ON)

L
Lag (G) – see clais
Làirig, learg & làrach (G)
Lanerc (P / B)
Law (ON / OE)
Lea, ley (OE) – see loan
Leac (G) – see clach
Lèana, lèanag & àilean (G) – see dail
Leitir, leachd, leathad & claon, faithir (G) – see also ruigh(e)
Links (OE / S)
Lios, lis, lus (G) & llys (B)
Llan, lann, len, lin, lon (B & G)
Llech, leac (B & G)
Llyn (B), linne & lòn, leàna / leànag (G)
Loan, loaning (OE / S) & lànaig / lonaig (G)
Loch (G)
Lùb, lùib (G) – see camas
Lwch, llaith & lleith (B) – see meg

M
Machair (G)
Maen or faen (B)
Maes, moss & cyn (B)
Magh, mauch, maich, mag (G) – see also maes
Maol & meall (G) & mol – see tràigh
Meg / mig & lwch, llaith & lleith (B)
Megin, mikla, minni & storr, peerie (ON)
Mersc, mersch, misk (OE / S)
Midden (ON)
Mire (ON) & moss
Mòine, mònach, mòinteach & monadh (G)
Moor, muir & moss (OE / S)
Muir, mara (G) – see tràigh

N
Nes(s), nis(h) & thang, hofud, horn (ON)
Neuk (OE / S)

O
Òb, oba, tòb (G) – see hop & vik
Oitir (G) – see tràigh

P
Pant (B)
Pasgell, pawr (P / B)
Pen, plen & pren (B)
Pert(h) (P / B)
Pevr (P / B)
Pit, pett(y) (P)
Pol, poll (B & G) & pol from bolstadr (ON)

Q
Quoy (ON) – see kyr & kvi

R
Ràth & caer, din / dun (B & G)
Raw(s) (OE / S)
Rhyn, rinn (B & G) – see ros & sròn
Riasg (G) – see bog
Rig, hrig (OE / S)
Rinn (B / G) – see sròn
Roinn(e), rann & raon (G) – see also earrann
Ros / rhos, rose (B & G) – see rhyn
Rubha (G)
Ruigh(e), ruidhe & rèidh, leitir (G)

S
Saetr, setr, shadder & svard, sword (ON)
Sàil, sàile, sàilean, sàilein (G) – see also tràigh
Scrog (OE / S)
Seggs, seggy (OE / S)
Shaw, schaw (OE / S)
Sheuch (S)
Sìthean, sliabh, slew & digh (G)
Skari, skarpr, sker, skeri & skal, skagr, sag (ON)
Sliabh, slèibhe (G)
Slochd, sloc (G) – see gleann
Srath, strath (G)
Sròn, strone (G) & trwn (B) & soc, corran, rinn (G)
Sta, stathir, setr, ster, from bolstadr (ON)
Stane, sten (OE / S)
Stav, stafr (ON)
Stede & stow (OE / S)

T
Taigh, tigh & ty (G)
Tairbeart, tarbet, tarbat (G)
Thang, tunga / tangi (ON), teanga (G)

Thwaite, twatt, thveit (ON)
Ting, ding (ON)
Tìr (G)
Tobar & fuaran (G)
Toll (G) – see clais
Ton / ington (OE / S)
Tòrr (G)
Tràigh, tràghad & gaineamh, muir, cuan, cladach, mol, bàgh, sàilean, dòirlinn, oitir (G)
Tref, tra, tree, try (B)
Tulach / tulloch, tull, tilly (G)

U
Uamh, uamha, uaimhe (G)
Ùidh (G) – see inbhir
Uisg(e) & bùrn (G), usc (B) & eas (G)

V
Vag, voe & hope (ON)
Vik, vig, wick, aig, uig, wall, way, voe (ON)
Vollr, wall, well (ON)

W
Wic, wick (OE / S)
Worth (OE / S)

Y
Yett, gate (OE / S)

6. PRE-CELTIC / EARLY CELTIC

The arrival of Celtic-speaking tribes on our shores is usually dated to around 1000 BC, but due to trade, especially in metal goods, it may go back much further than that. Before Celtic-speaking people arrived, we really do not know what language was spoken here, but we do know that it must have existed because of the ancient river names, especially the larger ones, which often cannot be explained, or explained satisfactorily, in any of the other languages known to us. These oldest names on our map, rather like linguistic fossils, come from our earliest settlers and have probably survived for thousands of years, while smaller burn names are perhaps more likely to come from later settlers.

Although many river names are an enigma or mystery, links between various river names across Europe may stem from common pre-Celtic or early Celtic river names for water / flowing water, or in some cases from the god / goddess that was believed to inhabit them: e.g. Allan, Alun, Alne, Ale, Alness, Allander, Ala, Aller, Alma; Aray, Ayr, Aar, Adder, Adda, Ahre, Ohre, Odre; Cart, Carron, Carrant, Carad, Cheron; Sheil, Seille, Sella, Saja or Shin, Shannon, Sinn, Sienne; Tain, Tay, Tyne, Teviot, Teith, Taff, Thames and Tamar; Dee, Don, Doon and possibly Deveron, Devon; Earn, Rhine and Rhône, or Almond, Afton, Avon, Avignon, Averon and maybe Girvan and Irvine.

Some of these can be traced to Indo-European roots like *el / ol* meaning to flow and *al / ala*, water, elements which appear in many river names across Europe, while Tweed, Tay and Tyne, from the Brittonic root *tae*, *teu*, meaning strong or flowing, have been traced back to a Sanskrit word *tavas*, meaning surging or powerful. Several of our river words are probably derived from the pre-Celtic or early Celtic *aulana* (as in River Allan and Allan Water) or *denona / deuona / devona* (from *deua*, a god, and *ona*, meaning river), as in the Don and Doon, or from *il / eil* roots, as in the River and Strath Isla (Perthshire), and possibly the River Eye in the Borders (OE *ae*, running water).

Another root is the hypothetical *ara / ar* (Brittonic), possibly as in Naver, or *urr / ur*, as in the River Urr, which is also the Basque word for water. The Brittonic river word *abona*, Welsh *afon* and Gaelic *abhainn* (pron. ahvin), or the shortened form *abh* (as in Avoch and Afton) are probably rooted in some of these old words and the fact that similarly sounding river names can be found across Europe shows how old they are.

The name Clyde, for example, a Cumbric / Brittonic name, may go back to the Indo-European root *clut*, meaning water or washing or alternatively meaning warm and sheltered, while it has also been linked, though without much foundation, to *Cluid* or *Clouta*, a river goddess (as in the River Cluden and Lincluden in Dumfries), and the word is still *Clyd* in modern Welsh. It was called *Clota* by the Roman historian Tacitus, *Cloid* by the Strathclyde Brittons and *Cloithe* by St Columba's biographer, Adamnan, while an earlier Cumbric name for Glasgow was Cathures, possibly from *Caer Clud*, fortress on the Clyde. The Scottish Gaelic word for the River Clyde is Abhainn Chluaidh and the word *Clutha* has often been used poetically to refer to the Clyde. Glasgow's old river ferries were called the Cluthas, while it is also the name of a famous Glasgow pub, as well as places in Canada and South Africa and even a river in New Zealand, so it has flowed a very long way indeed over many centuries and countries!

The fact that many river names may have passed through a number of languages without changing very much perhaps suggests that our ancestors thought it would have been unwise or tempting fate to change them, even if they originally just described the nature of the river itself, but some of them may have been regarded as sacred places, associated with gods or spirits of some sort, e.g. various Lochty or Lochy places are believed to come from the old / Irish Gaelic *loch*, meaning black / dark and *dae* (Scots Gaelic *dia*), a goddess, expressing the pagan belief that streams flowed with the spirit of a dark goddess, while Anu, or Danu, or Don, a Celtic mother goddess of fertility, prosperity and wisdom (converted into St Anne by the early Christian church) is believed to have left her name on various sacred water places featuring the word *an* or *dan*, as in Annan, Annick, or Burnanne, or Don and Danube.

Another Celtic mythological king or deity of many things, Lugh, features in names across Europe, such as Lyon (Lugdunum in Latin) and Leyden (Lugdunum Batavorum), Carlisle (Luguvallum) and maybe even London (Londinium). He was a sort of Mercury figure, a messenger of the gods, god of the sun and light, as well as the harvest, from the Celtic word *loucos*, meaning bright or white and whose name features in the Gaelic word for August, Lùnastal, the Scots Lammas.

In Scotland, Lugh features in various Ayrshire names like Lugar, meaning the stream of Lugh (with *ar*, an old Celtic river suffix), the Lugton Water and village, as well as maybe Loudon Hill (from Lugdunon, the wee fort of Lug), as well as the River Luggie in Lanarkshire, or even Relugas in Moray, another fort of Lugh. It is even claimed that the Celtic word *vindos*, white (as in Vindolanda on Hadrian's Wall), corresponds to or morphed into the Gaelic name Fionn, as in Fionn mac Cumhail (Finn MacCool) which may just be another name for Lugh, or his reincarnation. Old King Lugh certainly flowed far and he's never stopped running!

Pictish – the language of Calgacus

It is difficult to be sure about the exact identity and language of the Picts: possibly the descendants of, or related to, the first people to inhabit Scotland permanently. Their homeland was in the northern and eastern half of the country, mainly in Inverness-shire, Aberdeenshire and Fife, though their territory was once greater, including some of the Inner Hebrides, the Northern Isles, Caithness and Sutherland which they lost to the Vikings in the ninth century AD, though they left some very old men behind: the Old Man of Hoy in Orkney and the Old Man of Stoer in Sutherland (thought to be from Brittonic / Pictish *alt maen*, meaning high rock).

The term 'Picti' seems to have come from a Roman writer around AD 400, and a later Latin name for their kingdom was Pictavia. While there is no direct evidence that they called themselves Picts, the Old Norse name *Pettar / Pettir*, the Anglo-Saxon names *Pehtas*, *Pihtas*, *Pythas*, or even the Brittonic name *Prydyn*, are close to the Roman Picti, though these names could have come from the Roman word, but there were also Pictones or Pictavi tribes in Gaul. They might have called themselves the Coritani or Qritani, *Cruithnich* in Gaelic, as their territory was referred to as Cruithentuath in the twelfth-century Book of Leinster. Picti (Latin) may mean 'the painted ones' (i.e. tattoos, probably a derogatory term) or the people of the patterns or designs, and they were certainly very artistic as we can see in the great carved stones and jewellery they have left behind.

Other than some obscure Ogham inscriptions on these Pictish stones, we have no written evidence of their language, but as far as we can judge, they might have spoken a Celtic language related to Gaulish, a continental Celtic language spoken by the Gauls of France (Asterix and friends) but also related to Brittonic / Cumbric and Cornish, though there were significant differences between the tongues. There may even have been two (or more) Pictish dialects, e.g. possibly a northern and southern Pictish, one an older pre-Celtic Indo-European language and the other closer to Brittonic. Yet there was probably some overlap between Pictish, Brittonic and Gaelic, making it easier to borrow from each other or develop hybrids, especially as Pictish came under increasing influence from Cumbric and Gaelic and probably assimilated a fair amount of the latter by the eighth and ninth centuries or even earlier.

The leader of the Caledonii (the term earlier Roman writers used for them) at the Battle of Mons Graupius in AD 84 was, according to the Roman historian Tacitus, a man called Calgacus, or Galcacus, thus making him the earliest named person in Scotland's history. St Columba (sixth century AD) is said to have converted the Pictish King Brude to Christianity with the help of an interpreter, and Kenneth MacAlpine, the king of the Scots, possibly united the kingdoms of Picts and Scots around AD 850, though it was his grandson, Donald II, who was first referred to as Rì Alban, i.e. King of Alba, or Scotland.

Landscape Features / Natural Elements

Aber = mouth of river, estuary, confluence, often applied to settlements located there, as in Aberdeen (mouth of the River Don), Abercorn, a horned confluence, from Brittonic *corniog*, meaning horned; Aberdour, a water mouth, from Brittonic / Gaelic *dobhar*, water, or Arbroath /*Aberbrothoch*, which possibly means mouth of the little filthy or turbulent stream, from Gaelic *brothaig / brothach*, or maybe from a Pictish word *brudaca*, meaning much the same, while Fochabers, a loch mouth, is from a Pictish-Brittonic word *fothach*, loch or wet area. Aber is also Cumbric and appears in many Welsh place-names, such as Abergavenney or Aberystwyth. See **Brittonic / Cumbric**.

Cardden = wood, copse or thicket, often used as a qualifying rather than a generic element and is also a Brittonic / Cumbric word, as in Kincardine, head of the wood, with Gaelic *ceann*, head, or Cardenden, wooded glen or valley of the thicket, with Old English / Scots *denu / den*, a dell, hollow; Cardross, a wooded moor or wooded promontory, from Cumbric *ros*, or Pluscarden, a wooded place or possibly estate, with Pictish / Brittonic *plus / plas*, a place. Sometimes it is shortened, as in Cardno, Aberdeenshire, with Gaelic *ach*, a place, pruned to *o*, or it has been transformed via a process of sound shifts from 'cardden' to 'grand', giving us Grandtully, a hill thicket, from Gaelic *tulach*, a hill, or in Fettercairn, a wooded slope, from old Gaelic *faithir / fothair*, a terraced slope and *cardden*, shortened to or mistaken for *cairn*.

Settlement Names

Pit / Pett(y) = share, portion, piece of land, place or village, a much disputed word, but it appears mainly in areas of Pictish settlement and the specific / describing element is usually Gaelic (plus a few Brittonic ones), probably coined during a Pictish-Gaelic bilingual period in the ninth and tenth centuries, though it may sometimes be a translation of an earlier Pictish word, leaving us with many Pictish-Gaelic hybrids, often referring to a physical feature of the land, person, animal or number. Pit is a cognate (related, sharing the same origins) of the Welsh and Cornish *peth*, Breton *pez* and also Low Latin (later or medieval, non-classical Latin) *petia*, a piece of land. Pit survived into Gaelic, but was often replaced by *baile*, sometimes as late as the eighteenth or nineteenth centuries, though at least three hundred of them have survived, such as:

Pettie / Petty, a place of shares and Pettymuk = pig's place, from Gaelic *muc*

Pitcairn = cairn part, from Gaelic *càrn*, *càirn*

Pitcalzean = portion of the wood, from *coillean*, of the wood

Pitcarmick = Cormac's share

Pitcon = maybe farm of the dog(s), from Gaelic *coin / chon* (of a dog or dogs), near Dalry, Ayrshire, though possibly Brittonic

Pitgobar = goat's place, from Gaelic *gobhar*

Pitkerro = quarter share, from Gaelic *ceathramh* (fourth) and Pitcoig, fifth share, from Gaelic *còig* (five), as in Coigach, a place of fifths

Pitlochry = stony part or place, from Gaelic *cloichreach*, by / with the stones, stepping-stones

Pitlour = the leper's share, from Gaelic *lobhar*, leper

Pitmilly = mill place, from Gaelic *muileann*, mill

Pitmurchie = Murd's share

Pitsligo = place of the shells, from Gaelic *sligeach*, shelly

Pittencrieff = tree part, from Gaelic *craoibhe*, of the tree

Pittendreich = fine or braw portion, from Gaelic *dreach* meaning aspect or beauty, the most common *pit* name, but not dreich as in dull and gloomy as it is sometimes thought to mean

Pittenweem = the cave place, from Gaelic *na h-uamha*, of the cave

Pittodrie = wooded part, from Gaelic *fhodraidh*, meaning by the wood

Other words used by the Picts are *lanerc*, *pert*, *pevr* and maybe *pawr*, but they are also found in Brittonic and in Welsh, so they are listed below.

7. BRITTONIC / BRYTHONIC OR CUMBRIC / CYMRIC

The language of King Arthur?
The earliest language of which we have written record is Brittonic / Brythonic, also sometimes referred to as Cymric or Cumbric, the language of the ancient Britons which was used over most of Britain before the Romans arrived and up until the Anglo-Saxon invasions of the fifth century AD when it retreated north and west. The term Cumbric, or North Brittonic, is used to distinguish the variety spoken in southern Scotland and northern England from that used further south. There was, however, more than one Brittonic dialect and kingdom in Scotland, including the Lothians (the territory of the Votadini, later called Gododdin), which is thought to be named after a Cumbric chief, while their kingdom of Strathclyde had its capital at Alclud, or Alt Clut, meaning rock of the Clyde, but called Dumbarton by the Gaels, which simply means 'the fort of the Britons', a kingdom later referred to as Cumbria. Its power was weakened by Viking attacks in the late ninth century and the kingdom of Strathclyde merged with that of the Scots in the eleventh century when the power of the Gaels extended into the south west. However, Brittonic / Cumbric was probably still used over parts of central and south west Scotland until about the eleventh century, though gradually absorbed into the Gaelic of the Lowlands.

The Britons have left their names across our landscape and settlements, names which have survived around one thousand years after their language died out in these places, as the words lived on in important names, even if their meaning was forgotten, e.g. Arran is most likely to be from the Brittonic word *aran* meaning a high place or eminence, which certainly fits the topography, unlike the Irish Gaelic *árainn*, meaning a kidney, as in the Aran Isles, though another theory is that it comes from the Irish Erainn, changed to Arainn, meaning the people of Dalriada.

They have also left their mark on other high places, such as the Trossachs, which are apparently cross hills, from the old Welsh word *tros*, meaning across, while the Lomond Hills in Fife, as well as Ben Lomond and Loch Lomond, may be from *llumon*, a beacon (as in many beacon hills where signal or warning fires were lit e.g. Tinto Hill, from Gaelic *teine*), though Loch Lomond could well be from *leamhan*, elm, especially as the River Leven flows out of it. The large stone at the head of Glen Falloch, north of Loch Lomond, is called 'rock of the Britons' and is believed to have been the boundary between three Celtic kingdoms: Dalriada, Pictland and Strathclyde.

Dumbarton has its fort of the Britons but there is also a Dumbretton and a West Bretton near Annan, while Edinburgh has Baberton, which is possibly the steading or settlement of the Britons. Bartonholm (an old mining village, near Irvine) was the meadow of the Britons, from ON and OE *holm* and Bartonhill, another old mining village near Baillieston, Glasgow, was once the Britons' hill, similar to Barbrethan in South Ayrshire, the ridge or the height of the British, while Auchenbart, near Darvel, was probably the field of the Britons. The town and bridge of Erskine could be from *ir ysgyn*, meaning a green ascent (though maybe also from Gaelic *ard sescenn / seasgann*, a high marsh) while Cowcaddens in Glasgow might have been a wooded hollow (see **cau** below). Possil in the same city was, at one time, a place of rest, as in the Old Welsh *poues*, meaning rest, repose, or stance.

Names that may have a Brittonic origin can be found across Scotland from the Borders, such as Peebles (sheilings or tents), to Central Scotland, such as Linlithgow (lake in the moist hollow), to the Pictland of the north east where various *Aber* places, including Aberdeen, are from this shared Brittonic / Pictish word, while Buchan might well just be a coo place (from Brittonic *buwch*, cow). Brechin might commemorate the same legendary Brittonic hero as Brecon in Wales and Llanbryde with its Welsh spelling, near Elgin, is Bride's church, though it could also be from Gaelic. No wonder that the character Jerry Melford, in Tobias Smollett's great eighteenth-century novel *Humphry Clinker*, remarked on the 'great affinity' between 'Gallick' and 'Welch', something he hears in their speech and sees in the people and the landscape.

The Island of Cumbrae is simply another name for the Brittonic / Cumbric speaking people of Strathclyde, very similar to Cymru or Wales (as in Plaid Cymru, Welsh National Party) or the district of Cumbria in north west England. Their legacy also survives in the surname Galbraith, from Gaelic 'Gall Bretnach', meaning stranger Briton, while the legends of King Arthur and Merlin belong to this period and there is no shortage of places associated with his legend in southern Scotland, as in various Arthur's Seats (Edinburgh and Borders), Banff in Aberdeenshire, or Loch Arthur, Dumfries, while some think the Isle of Arran is the legendary Isle of Avalon, the magic apple island of regeneration and youth.

Brittonic was a Celtic language, the ancestor of old Welsh, known as P-Celtic, as opposed to Irish and Scots Gaelic (often referred to as Goidelic) which are Q-Celtic, because P-Celtic often has a 'p' sound where Q-Celtic has a 'kw / qu' sound, usually a hard 'c' in modern Scots Gaelic e.g. in modern Welsh the word for head is *pen* and in Scottish Gaelic *ceann* (pron. keown), the Welsh word *plant* for children is *clann* (clan) in Gaelic, while *map* for son is *mac* in Gaelic, though both Celtic branches shared many words, just as Gaelic and Welsh still do.

Landscape Features / Natural Elements

Aber = mouth of river, estuary, confluence, a word shared with the Picts, e.g. Aberfeldy, possibly meaning river mouth of Phaldoc or Paldoc, an early Christian monk, or from Peallaidh, thought to be the name of a local water sprite, and the rivers were full of them at that time. Aberfoyle is a confluence of the pool or mouth of the streams, from Gaelic *phuill*, of the pool, Aberlemno is mouth of the elm stream, from Gaelic *leamhan*, elm, Abernethy is the confluence of rivers Nethy and Earn, maybe from a Brittonic river name, Nedd, though it is uncertain. Other examples: Aberlady, Abertarf, Fochabers, Kinnaber or Lochaber.

Al / ail = rock, rocky, a Brittonic-Pictish word that passed into old Gaelic, as in Ailsa, Alloa, Alloway and Alva, from *ail-mhagh*, a rocky plain, or Alvie, a rocky place. See also **maen**. The Brittonic and Welsh word *carreg*, a rock or cliff, also passed into Gaelic. See **carraig**.

Blaen = edge, valley head, upland, or a river source in Welsh place-names, found in Blantyre and possibly Blanefield and Strathblane in Stirling, or Mossblown, Ayrshire (from *maes*, plain, field), though Dunblane is thought to be from St Blane, but may well be from *blaen*, a word which could also indicate a settlement.

Cardden = see Pictish section, p. 40.

Cau = hollow, dell, as in Glasgow (*glas*, meaning green or grey), as discussed in the introduction, which could perhaps be from *cu*, meaning dear, but, like many things in Glasgow, still argued over; Linlithgow (see **llyn** below), Cowcaddens could be from *cau* and *cardden*, meaning wooded hollow, though it is usually given as hazel nook, from Gaelic *cùil*, corner, nook, with *calltainn*, of hazels. It maybe also features in Moscow, Ayrshire (see **maes** in settlement section). A related word is **coomb / cwm**, a shallow valley, found in Welsh names, as in Cwm Rhondda, the name of a former coal-mining valley and a famous hymn tune.

Cefn = ridge, as in Giffen in North Ayrshire, or Giffnock in East Renfrewshire, a little hill, from Cumbric *oc*, little, or possibly in Govan (see **maen**) and maybe even Cheviot. **Cyffin** also means a boundary or frontier in Welsh names.

Coed / coet (*coed* or *goed* in Welsh) = wood, forest, as in Pencaitland (see below), but it sometimes became gate / gait (way, road) as in Bathgate, which is not a bath gate but is from *baedd coed*, boar wood, or from *bod*, a house, so maybe it was a house at or by the wood, or a house in the wood that had a pig, but nobody can now remember! Dalkeith is definitely a wooded field or meadow, from *dol coed*, but Keith is just a wood, from *coed* via old Gaelic *coit*.

Dobhar / dwr (*dwr* or *dwfr* in Welsh) = water, stream (pron. doh-wur), from the same root as Dover and found in Aberdour, mouth / estuary of the stream, Calder and Cawdor, possibly meaning hard or rapid water, from *caled*, hard, or Edradour, between two waters, from Gaelic *eadar*, between, and *dà*, two, or Condorrat, a river confluence place, from Gaelic *comh*, joining, *dobhar* and *àite*, a place. Dobhar was also Pictish and passed into old Gaelic, as in Ben Dorain, from *dòbhrainn*, of the streams, or Darvel, water of the town, from *dobhar* and *bhaile*.

Llech = flat stone, found in various Welsh names, or in Wanlockhead, from *gwyn / wyn*, white, with head added much later to the old name Wenlec, the name of the nearby river which obviously had a lot of white flat stones. *Leac* is the Gaelic equivalent, as in Auchinleck. See Gaelic section.

Llyn = lake, pool, waterfall, as in Lindifferon, from *llyn diffryn*, valley lake, Lincluden, pool on the river Cluden, Lindores, from *llyn dwr*, water lake (possibly another pleonastic name), or alternatively a dark wood lake from Gaelic *dhu* and *ros*, or Linlithgow, pool in the moist hollow, from *lleith*, moist, and *cau*, hollow, or as in the obvious Linwood. See Gaelic **linne**.

Maen (or sometimes mutated to **faen**, *maen* in Welsh, *men* in Cornish) = rock, stone, e.g. Methven, near Perth, middle stone or boundary marker, from Cumbric *medd* and *faen*, or Manuel in West Lothian, from *gwel*, view, outlook, or possibly the Penmanshiel Tunnel in the Borders, from Cumbric *pen*, head, *maen* and Norse *skali* or Scots *shiel*, a sheiling or shelter. See also **al / ail**.

Meg / mig (also used by the Picts; *mign* is a bog in Welsh, and maybe the source of the Scots *mingin*, stinking!) = mist, drizzle, water, bog, swamp as in Meigle in Perthshire and a hill in the Borders, Meggat Water in Dumfries and Galloway, the River Meig in the Highlands and Strathmiglo in Fife, a swampy valley, or maybe Meigle Bay, North Ayrshire, but Megginch is from old Gaelic *melg*, milk. Other soggy words are **lwch**, marshland, maybe as in Carluke, fort on the marsh, similar to Gaelic *luachair*, rushes, **llaith**, moss, bog, and **lleith**, a wet place, as in Leith where people often take a while to dry out.

Pant = hollow, valley bottom, maybe as in Pinkie Cleugh, near Musselburgh (site of a famous battle in 1547), a wedge-shaped hollow, from *pant* and *cyn*, a wedge, or Panmure, a big hollow, from *pant* and *mawr*, big, both words used in Welsh, meaning big hollow or big bottom and there is no shortage of these today!

Pen = head, top, summit, headland, chief, and also source of stream, found mainly in south-eastern parts and around the Solway (and very common in Wales), usually referring to hills and can appear as *pin* or *pun*, maybe as in Pinkie (see **pant** above), or as *pen* in Pentland Hills, Pencaitland, meaning head or hill of the wood of the enclosure, from *coet* / *coed* and *llan* / *lann*, field, enclosure or Penicuik, hill of the cuckoo, from *y cog*, the cuckoo, or in Penmanshiel in the Borders, or used with a prefix, as in Cockpen, meaning red hill head, from Brittonic *coch*, red (borrowed from Latin *coccum*, scarlet berry). It possibly also appears in a few North Ayrshire places, e.g. Pinnioch Point, Dalry, possibly meaning little summit and Pundeavon, river head or source, in Kilbirnie. Pen / pin is sometimes confused with the Gaelic *peighinn*, penny, a land rental value, as in Pinmore, the big penny land. **Plen**, though, means flat land, or a plain, as in Plean in Stirlingshire.

Pert(h) (perth in Welsh) = bush, thicket, wood, as in Perth, Larbert, meaning ruin by the wood, from Gaelic *làrach* and *pert*, and in Partick, from *perth* and *aig*, little, so a little wood. Also Pictish and in Pictish areas it was elevated from a natural feature to a settlement word.

Pevr (*pefr* in Welsh) = radiant, shining, beautiful, mainly used for stream names, as in Strathpeffer, a bright valley, Inverpeffer and Inverpeffrey, shining river mouths, Peforyn, now Silver Burn in Aberdeen, two Peffers in East Lothian and Peffer Mill in Midlothian.

Pol = pool, bog, mire, flowing water, burn, hollow (see Gaelic **poll**), e.g. Polmaddie Hill in South Ayrshire, probably from *pol* and *madadh* / *madaidh*, hound(s), of dog(s), so pool of the dogs (see Gaelic Polmadie) and Pollokshaws is a little pool of the wood, while Pollokshields is a wee pool of the hut or sheiling. Also *pwll*, as in Welsh, which appears in Carpow and various Pow Burns in Scotland.

Pren (also Welsh, and *crann* in Gaelic) = tree or wood, but some prens evolved into settlement names, as in Traprain in East Lothian, homestead of the tree, from *tref* and *pren* (see **tref** below), Primrose in Fife (and a family name), meaning tree of the moor, from *ros*, moor or headland, or Barnbougle on Dalmeny Estate, West Lothian, derived via several sound shifts from Pronbugail (*pren* and *bugail*, a herdsman, *buachaille* in Gaelic), so meaning the herdsman's tree and, via more examples of metathesis, it appears in various Pirn or Pirnie places, and even in prim places like Primside, Roxburgh.

Ros / rhos (as in Welsh and also Gaelic) = promontory, peninsula, isthmus, point or moor or wooded headland (can also mean wood in Gaelic), as in Melrose, from *Moelros*, a bare moor, Roslin which probably refers to a holly moor from Cumbric *celyn*, holly, while Culross is a holly point or wood, from Cumbric *celyn* or Gaelic *cuileann* and both Dores on Loch Ness and Durris in Aberdeenshire are dark woods, from Gaelic *dubh* and *ros*, though Rosyth, on the Firth of Forth, is either from English *hide*, a landing place, or Gaelic *saighead*, an arrow. Ros can also appear as *rose*, as in Rosemarkie, from Gaelic *marc*, a horse or steed. **Rhyn** (*rinn* in Gaelic) is another Brittonic word for point or promontory, maybe as in the Rinns of Galloway or Islay, or Renfrew, point of the current, from Brittonic *frwd*, a current.

Usc = water (*uisge* in Gaelic), as in Throsk, near Bannockburn, from *tref* and *usc*, so a house at or near the water, i.e. the Forth, while in Ayrshire there are Dusk and Duisk Burns, both dark waters maybe from G *dubh*, dark, and *uisge*, or from the earlier Br equivalents, *dwr* and *usc*.

Settlement Names

(i) **Caer / cair / car** = fort (also Gaelic) and can appear as **cader** or **cathair** (seat or fortified place) in Gaelic, as in Stracathro, in Angus, a valley fort, from Gaelic *cathrach*, of the fort, but we have them all over the place in various forms, as a Gaelic suffix or prefix. Caerlanrig in the Borders is a fort on the glade, from *llanerc*, a forest glade, while Caerlaverock, near Dumfries, or Carlaverock near Tranent, were possibly forts named after a Brittonic king called Llywarch, but later influenced by their similarity to the OE / Scots word *laverock*, a lark, though also claimed to be from the Gaelic *leamh-reaich*, of the elms. Other examples are Cardonald, Donald's fort, Carriden, fort on the slope, from *eiddyn*, slope / face, Castlecary, castle fort, with English castle overlapping the older *caer*, Catterline, fort by the pool, from *cader* with Brittonic *llyn* or Carpow, the fort on the pool, from Pictish / Brittonic *pwll* / *pol*, a pool, or Cramond, fort on the River Almond, another ancient pre-Celtic river name.

(ii) **Caer**, a fort, is often changed to **kir, ker** or **kirk** as in Kirkcaldy, fort on the hard hill, from *caer caled din* and this is probably what happened with Kirkmichael in South Ayrshire, as older folk still call it Carmichel. It can also appear as *quhar*, as in Sanquhar, from *sean caer* or *seann chathair*, old fort / fortified place, seat or throne. Some *car* words are maybe from an early Celtic root *kar*, meaning rough, harsh (also the root of *càrn*, a cairn), as in Carrick (Brittonic *carreg*, Gaelic *carraig*, a rock or rocky place), or Carron and Loch Carron. Other examples: Carlisle in England, Cardiff, Caernarvon and Carmarthen in Wales, Cardinham and Kerrow in Cornwall. The Brittonic-Pictish *cader* appears in Cadder, Dunbartonshire, and becomes *cathair* in Gaelic, a chair, throne, city or fort, as in Catterline, Aberdeenshire, a fort by the pool (Gaelic *linne*). See also **ràth** below.

Din, dun, dum (Welsh *din* and Gaelic *dùn*) = fort or hill, as in Gordon or Gourdon, both great forts, from *gor*, meaning great, or Dumbarton, fort of the Britons, or Dunipace, near Falkirk, hill of the pass, from the Cumbric *din y pas*, but *din* has become *tan* in Tantallon, East Lothian, with *talgan*, a high front or brow, while Tentsmuir in Fife has maybe evolved from *dinas*, a fort place, with the Scots *muir* added later. Yet some duns in Lowland Scotland could also be from Old English *dun*, a hill or rise (via Old Norse / Danish), such as Granton, a green hill in Edinburgh, or Loudoun Hill in Ayrshire, maybe with the Scots *lowe*, fire, via Norse *logr*, or alternatively from Lugdunon, the wee fort of Lug. (See Lugh in Pre-Celtic / Early Celtic section, p. 37.)

Dol (see Gaelic **dail** / **dal**) = a field or meadow, as in Glen Doll or Dollar, Clackmannanshire, and probably Dallars, Ayrshire, from *dol* and *àr*, meaning ploughed or arable fields.

Egles = a church (Welsh *eglwys*, Cornish *eglos*, Gaelic *eaglais*). Most of our Eccles or Eglis names originate from Brittonic rather than Gaelic, e.g. Ecclefechan, meaning the church of Fechin, or alternatively little church, from the Cumbric *egles-bychan*, or Ecclesmachan, the church of St Machan or Eaglesham, just meaning church village, or Terregles, another village of the church, from Brittonic *tref* and *egles*, or Gleneagles, glen of the church (not glen of the eagles), which is from Gaelic.

Gafael = a smallholding, possibly as in Dungavel, fort of the holding, or Mossgiel, field of the holding, from *maes*, a plain, open field. See **maes**, below. It means holding, grip or tenure in Welsh, and appears in Gaelic as *gabhail*, though sometimes confused with *gobhal*, forked, maybe as in Dungavel.

Lanerc (*llanerch* in Welsh) = a forest glade or clearing, as in Lanark, Lanrick in Perthshire and Lendricks in Angus and Stirling, Caerlanrig near Roxburgh, or in Barlanark, either from *baed*, a boar or from Gaelic *bàrr*, a ridge. Also Pictish.

Llan / **lann** = field, enclosure, house, church or parish, very common in Wales, though *llan*, as in Llanbryde in Moray, is very unusual in Scotland. It often appears as **len, lin, lon, lum, long**, all of which can mean various things! Thus we have long places like Longannet, field or church of the patron saint, from the Gaelic *annaid*, Longforgan, a boggy field / church (from Gaelic *gronn*, a marsh), Longmorn, not a long morning but Morgan's field or church, Longformacus, Maccus's church, or various **lumph** places like Lumphanan in Aberdeenshire and Lumphinnans in Fife, which mean the land of St Finnian, a church dedicated to him, from Gaelic *lann*.

Llys = garden, enclosure, which appears in Leslie, pool garden, from *llys* or Gaelic *lios*, garden, and *linn*, pool (with the *nn* dropped), or maybe holly garden, a contraction from *llys* and *celyn*, holly (with the *n* dropped) though it may also be derived from Pictish, as it was a word they had in common.

Maes = plain, open field (the same in Welsh), as in Mossgiel, Ayrshire (where Robert Burns once lived), field of the smallholding (see **gafael** above), but later changed to *moss* and the *f* of *gafael* dropped, or the nearby Mossblown, from *maes* and *blaen*, edge, top. Also from the same root, we have Menstrie in Clackmannanshire (see **tref**) and also probably Moscow, an Ayrshire field in a hollow (Br *cau*), which was renamed after the Russian capital city to celebrate Napolean's defeat in Russia in 1812. Maes is also common in Welsh names. Another Cumbric plain word is *fa*, as in Ogilvie, meaning high field, from *ocel fa*. **Cyn** is another Cumbric field word, as in Gorgie, from *gor*, big, plus *cyn*, while *dol* was also used (see above). Also found in old Gaelic (see **dal**).

Pasgell / pawr = pasture, grazing, possibly as in Paisley, meaning pasture slope, from *pasgell* and *llethr*, a slope, or Balfour, a settlement on a pasture, from Gaelic *baile* and *phuir*, the genitive of *por*, pasture, but borrowed from Pictish or Cumbric and still found in modern Welsh as *pawr*.

Ràth (also old Gaelic, pron. rà) is another old name for a ringed fort, broch or residence, e.g. Dounreay, Reay, Rothes, Rothiemurchus and Ratagan, a wee wee fort (all in Highland Region), Rattray (see **tref** below) and Rohallion, fort of the Caledonians in Perthsire, Ratho in West Lothian, Romanno in the Borders from *manaich*, of a monk, or Rathillet in Fife (maybe fort of the Ulster man / men, from G *Ulaid* or *Ulad*) and Raith Rovers, the football team from Kirkcaldy, though Glenrothes is a modern name.

Tref is a very old word (found all over Wales and Cornwall) for a farm steading, settlement, or village, appearing as the first or second element in a word, and is written in various forms: **tra, trie, try, tray, trab, trap, tre, tree, drie** and also as *tref / dref* or *tre / dre* in Wales. It is much less common for the generic, or classifying, element to appear second in Gaelic names, though more common in Brittonic ones. A Cumbric first element usually appears first in the south, but mostly a Gaelic one in the north, as in Fintry, meaning white farm, from *fionn tref*, Menstrie which is either from *maen*, a stony dwelling, or from *maes tref*, a plain dwelling, or Niddrie, a new dwelling, from *newydd tref*, and Longniddrie, church of the new dwelling (from *lann* and *newydd*, new and *tref*). Ochiltree is a hill or high farm, from *ocel*, high, as in the Ochil Hills, Rattray is a fort farm, from *rath*, a circular fort, and Soutra is a farm with a wide view, an outlook house, from *sulw tref*. South of the Forth-Clyde line, the *tref* root usually comes first as in Throsk, house by the river, from Cumbric *usc*, water, Trabrown, hill village, from *yr bryn*, of the hill, Traprain, tree house or homestead of the tree, from *pren*, a tree, or Trabboch in Ayrshire from *tref* and G *bac*, a hollow, bend, crook, or bank and Tranent, village by the valley or stream, from *tref yr neint*. Threave is probably just a settlement, but could be from the Gaelic *treabh*, to plow or till, while Traquair is just a farm on the River Quair, an odd name right enough! Trabboch (near Old Dailly, Ayrshire) is also probably from *tref* and G *bac*, a hollow, bend, crook, or bank. One possible exception to this pattern is Belltrees, near Lochwinnoch, with the G *baile*, village, added to the older Br *tref / trees* word.

8. GAELIC – THE LANGUAGE OF THE FIRST 'SCOTS'

Gaelic is often said to have arrived in Scotland from Ireland with the Scots (from 'Scoti', meaning raiders, another name given to us by the Romans) in the late fifth century, but it is now generally accepted that there was a lot of movement between Scotland and Ireland before that period along the western coastal highways, with Gaelic probably being used in parts of Scotland from earlier times. Yet, once the Scots expanded their kingdom of Dalriada from Ireland into Argyll, they gradually extended their influence into other parts of Scotland, a process that eventually led to Gaelic mingling with and eventually replacing Pictish and Brittonic from around the tenth and eleventh centuries. However, these two Celtic languages had much in common and there must have been a lot of linguistic plurality or co-existence involved, i.e. more than one language spoken side by side, or people even speaking a mixture of the two for some time.

Kenneth MacAlpine and his successors united the kingdoms of the Picts and Scots in the second half of the ninth century AD and Gaelic thus became the language of power and prestige, something that had a huge influence on our culture and place-names. Gaelic came to be used over most of Scotland for the next five or six centuries, though with a thinner and more temporary presence in the south east, or roughly north and east of a line from Dumfries to around Musselburgh, where Anglic was already established. Yet, by the twelfth and thirteenth centuries, the Norman French of the nobility and the northern English / Northumbrian of the commoners had made deep inroads into the Gaelic-speaking areas, though most Scots probably still used or understood Gaelic in some form until around the fourteenth or fifteenth century. By the late fifteenth century Gaelic was mainly used north of the Highland line, though it continued to be used in some places, including the Lowlands, until much later, e.g. Perthshire, Galloway, Kintyre, Arran and parts of Ayrshire which are all rich in Gaelic or Brittonic names.

Gaelic also suffered increasing political and cultural hostility from the Scottish kings and ruling classes, firstly with the growth of Norman influence and power from the eleventh and twelfth centuries onwards, secondly with Stuart monarchs trying to break the power of the Lords of the Isles and then the Kirk taking an increasingly disapproving view of Highland Catholics and Episcopalians, after the Scottish Reformation of 1560. Thus, over the centuries, Gaels came to be seen as a threat to the central power of church and state, with Gaelic looked upon as barbaric and uncivilised, as shown by the Statutes of Iona, 1609, which decreed that all Highland chiefs send their sons to English-speaking Protestant schools in the Lowlands.

Hostility towards Gaelic increased greatly after the 1745 Jacobite Rising, with the Proscription Acts aimed at destroying the clan system and disarming the Highlands. Measures like this were rooted in a hostility that continued throughout the nineteenth century with various Education Acts which have left a long legacy of discrimination towards Gaelic that has lasted well into the twentieth century and is still sometimes voiced today by people ignorant about Gaelic history and culture.

For students who live in the Highlands or Western Isles, it should be very easy to identify Gaelic names, not only for landscape and settlement or farming words, but for birds, animals and plants, while other parts of the country are not short of Celtic names, both Brittonic and Gaelic, especially the south-west where the landscape is covered in them. A first step would be to simply check the map to identify some of

Exploring Scottish Place-Names 57

these names, even in towns and cities of the Lowlands where many of the very Scottish-sounding names are from Gaelic. Keep in mind too that Gaelic, like our other languages, has a number of dialects and so the same or a similar word can sometimes have various meanings.

Grave accents, indicating long vowels, have been inserted in **head words** and where Gaelic words or names are quoted, but not in names as they usually appear on the map. Genitives / possessives are often also given as they are very common in Gaelic place-names.

Landscape and Coastal Features

Abhainn (pron. ahvin / awin), **aibhne** (gen.), a cognate of Welsh *afon* = usually refers to a larger river flowing into the sea, with many tributaries, and is a word connected to ancient European river names from Irvine to Avon and Avignon, as discussed in the Pre-Celtic section, e.g. Loch and River Avon in Moray or Ruthven in Highlands and Perthshire, places with red rivers, from Gaelic *ruadh*, red; Strathhaven, a river valley in Lanarkshire, and from the shorter old Gaelic form **abh**, meaning water or a stream, we have Avoch in the Highlands and Afton in Ayrshire. The River Girvan in that shire might mean short or rough river, from Cumbric *gerw* (as in G *garbh*) and *afon*, river, or Gaelic *geàrr*, short, but it is much debated, and is now thought by scholars to be either from old Gaelic *gar*, a thicket, and *fionn*, fair, white. Irvine, also much debated, is possibly either from Gaelic *iar*, west, and *abhainn*, or the older Brittonic *yr*, the, and *gwyn*, white, so maybe it was the white or west-flowing river.

Allt (pron. ault) = burn, stream, especially with a steep bank, very common in the Highlands, while *alltan* is a small burn or rivulet, though *alt* (Cumbric / Welsh *allt*) can also refer to a cliff or height. They flow all over, e.g. Altnabreac, trout burn, from *breac*, meaning speckled, as in a trout's skin, Aultbea, birch burn from *beithe*, birch tree, Taynuilt, a house on the stream, from *uillt*, the genitive / possessive of *allt*. Allt is sometimes corrupted into English Old or Scots Auld or Ault, as in Old Maud from *allt madaidh*, dog's burn, or Auldearn, stream of the Earn or Auldbar, stream of the height, from *bàrr*, ridge or height, and Auldgirth, from *gart*, field. 'Auld Reekie' is the old nickname for Edinburgh, once a very smoky and smelly place, though it is probably from Gaelic, *alt-ruighe*, a high slope, which certainly fits the topography. This latter **alt** means a height or cliff, and, like the Welsh *allt*, is from the Latin *altus*, high, as in altitude.

Aonach (pron. oenuch), **aonaich** (gen.) = steep hill, ridge, plateau, slope, or moor, as in Aonach Eagach, the notched ridge in Glen Coe, Aonach Beag and Aonach More, the small and large ones. We also have **aoineadh**, a steep rocky brae or promontory.

Àrd, àird = high place or promontory, as in Ardchattan, height of St Chattan, another Celtic monk, or Ardeer, west cape or headland, and Ardersier, with *ros*, a point, plus *iar*, west. Ardentinny means height of the fire, from *teine*, fire, i.e. a beacon or signal fire, though *àird an t-sionnaich*, height of the fox, is also a possibility (as with Craigentinny), while Ardkinglas is height of the dog stream, from *con*, dog, and *glas*, water; Ardnamurchan is height of the otters, from *na muir-chonn*, of the sea dogs or otters (though it could also refer to pirates), while Ardmore is a big (*mòr*) promontory, almost as common as Airdrie or Airthrey, both high hill slopes, from *ruighe*, slope, or Ardoch, another high place or land, with Gaelic suffix *ach*, land. Ardrishaig is a prickly place because it is the height of the thorns, from *drisean*, *driseag*, brambles or little thorns, while Ardrossan really stands out as it is the height of the point or promontory, from *rosan*, little headland (a diminutive form of *ros*), but an *àrdan* is just a hillock or any piece of rising ground. Some *àrd* or *art* places are, however, from the Norse *fjordr*, a sea loch or bay, as in Minard or Moidart. See Norse section.

Àth (pron. ah) = ford, as in Alford, a ford ford, with the English ford superfluous, or in Loch, Glen and the River Affric, from *breac*, trout or *bhraich*, of a boar, or Atholl, possibly ford of Fodla or new Ireland, as *ath* also means new, next; Aboyne, meaning the white cattle ford, from *àth* and *bò* (cattle) and *fhionn* (white); Ibrox, ford of the badgers, from Gaelic *bruic* or Scots *brock*, but well hidden in Acharacle, Torquil / Torcuil's ford, and even more so in Amulree in Perthshire, the ford of one Maolrubha. **Fadhail** is another ford (or strand) word, as in Benbecula, Beinn nam Fadhla (or Faoghla) in Gaelic, i.e. hill of the ford.

Bad, badan = spot, place, area or tuft, cluster, clump, thicket and we are not short of bad places, from Badbea, a birch spot, from *beith*, Badcall or Baudcall, hazel places, from *coll*, though Badenoch is from *bàthte*, drowned, flooded, and *ach*, a field, place. See also **coille**.

Bàgh, bàigh (gen.) (pron. baagh / biye) = a bay, as in Bàgh a Tuath, north bay, Bàgh Mòr, the big bay, Bàgh a' Chaisteil, Castlebay (see **camas** and **traigh** below).

Bàrr = height, top, summit, ridge, as in Bardowie, a dark height, from *dubh*, dark, or Barrhead, a top head or ridge, or Barlinnie, a pool on the summit or high pool, from *linne*, pool, though today more famous for its jail. Dunbar is a fort on the height, from *dùn*, while Longbar, Ayrshire and elsewhere, is maybe from Old Gaelic (Irish) *long* which has several meanings, with encampment or field being the most likely, or it could even be from Br / G *lann*, enclosure.

Bealach = pass or gap, usually between hills (*bwlch* in Welsh) as in Bealach na Bà (of the cow) and Bealach nam Bò (of the cattle), or Tayvallich in Argyll, house in the pass, from Gaelic *tigh* and *a' bhealaich*, of the pass, or Ballochmyle in Ayrshire, from *bealach* and *maol*, brow or bare. Balloch on Loch Lomond refers to the river pass or gap from the loch to the Clyde via the River Leven.

Beinn (pron. byn), usually appearing as **ben** = hill, mountain or peak (and the genitive plural **beann**, as in *tir nam beann*, land of the bens), usually the highest elevation in an area and we have about one thousand of them, including our highest one Ben Nevis, whose name is debateable, possibly from the pre-Celtic root *nebh* (cloud, water, mist) as is the name of the river and loch of the same name. *Binnean* is a diminutive (smaller) form, often meaning a point or pinnacle, and both words are usually qualified or described in terms of size, colour, shape or form, as in many called Beinn Mhòr / Ben More, big hills, but there are some really small ones like Ben Venue, from *meanbh*, tiny, though size is often relative and we have mountains of many colours or shades, which often depend on perspective or distance, especially *bàn* / *bhàn* (white, light, fair), *breac* / *bhreachaidh* (speckled), as in Ben Vrackie, *buidhe* (yellow), *dubh* (dark / black), *dearg* (red), *donn* (brown), *geal* (bright, shining), *gorm* (blue or green), *glas* (green or grey), *liath* (grey), *uaine* (green). See also **monadh** and other mountain words below.

Bog = soft, damp, miry, boggy, and **bogach** (pron. BOGEuch) or **boglach** (pron. BAWkluch) both refer to a marsh or bog, as in Bog of Gight, Aberdeenshire, a windy bog (from Gaelic *gaoithe*). Scotland is covered with thousands of bogs and there are over forty different names for bog in Gaelic, surely proof, if needed, that we are one of the soggiest places on earth! **Riasg** or **rèisg** (gen.) is one of these bog words that appears in many Risk places, especially farms, while **fèith**, a boggy place, appears in the Glen and River Feshie in the Highlands. See **innis**.

Bonn (pron. bounn) and **bun** = bottom, base, root or mouth of river, the opposite of ceann / ken, as in Bunnahabhain, Islay, mouth of the *abhainn*, river, or Bunessan on Mull, foot of the waterfall, from *easan*, a wee waterfall, while there is even a Boon in the Borders.

Bràigh (pron. bry) = brae, slope, upper part, uplands, from Norse *bra*, brae, as in Braeriach, in the Cairngorms, from *riabhach*, brindled, striped or greyish, Braemar (from a personal name) or Braemore, the big upland. Other sloping words are **bruthach**, a steep hill, brae or ascent, and **bruach, bruaich** (gen.), a bank or brim, as in Breich, West Lothian, or Tighnabruaich in Argyll, the house of the bank, but not the bank house, or Bruach na Frìthe in the Cuillins on Skye, bank of the deer forest. **Bràghad** is another Gaelic word for an upper part or hill area, sometimes changed to the Scots word *braid*, broad, referring to upper slopes, as in Braid Hills, Edinburgh or Breadalbane, in Perth and Stirling.

Camas, camus or **cambus** = bay, channel, creek, or crooked rivulet, as in Camus Beag (small bay) or Cambuslang, a ship creek on the Clyde, though inland *camas* means a bend, as in Campsie Fells, and surrounding Campsie places, from **cam, chaim** (an old Celtic word), meaning bent or crooked, plus *sith*, knoll, hillock (see below) and the later Norse Fjall / Fell, while Camelon is also from *cam* and *linne*, a pool, but nothing to do with Camelot and King Arthur! Other bent or crooked words are **lùb, lùib**, a bend, as in Luib or Loch Lubnaig, meaning wee bent loch and **crom**, as in Cromarty, crooked height, from *àrd*, height, Cromalt, a winding burn and Crombie, probably also referring to a winding stream, but not to hats or coats! Ancrum in the Borders is from the Cumbric cognate *an crwm*, just meaning the bend.

Caol, caolas (pron. kule, kulas), **chaolais** (gen.) = strait, narrow stretch of water, channel, sound, usually written as **kyle**

as in Kyles of Bute or Kyle of Lochalsh, though Kyle in Ayrshire, supposedly named after a fifth-century tribal ruler, Coel Hen (old), as in Old King Cole, may just have come from *coille*, a wood. We have them all round the west coast and the islands, as in Ballachulish, village of the narrows, from *baile chaolais*, Kyleakin, straits of Hakon, the Norse King at the Battle of Largs in 1263, Kylerhea, straits of Reathan, a hero in the Ossianic myths about Fionn MacCumhaill (Finn McCool), a really cool guy, or in Kylesku, a narrow strait and definitely not squinty, from *caolas* and *cumhann* / *cumhang*, narrow. There is also sometimes confusion with **cùil**, a nook, corner, or *cul*, a back place (see below), as in Culroy (not Kilroy), *cùil* and *roy*, from *ruadh*, red, near Minishant, Ayrshire and sometimes it is even confused with *coll* / *calltain* = hazel(s). Other variants or related words are **cumhang** (pron. coo-ung), cung, cuinge, a narrow strait channel or defile, and other disguised forms of *caol* are *col*, *coe*, *chu* or *kil*, such as Colintraive, the swimming strait, from *caol an t-snàimh* (where cattle used to swim across), Glencoe, narrow glen, or Kilchurn, straits of the cairn (and name of castle at the narrow part of Loch Awe), or Eddrachillis, between two channels, from *eadar dhà chaolas*. See also **cille** / **kil**.

Càrn / **càirn**, **cùirn** (gen.) = a humped hill or heap of stones, a common prefix to mountain or hill names, as in Cairndow, black mount (from *dubh*), Cairn Gorm, blue humped hill, from *gorm*, blue, or Cairnpapple in West Lothian, pebble hill or cairn, from Old English *popel*, a pebble, or alternatively, hill of the priest's place (Gaelic *pap* and *ail*, location or place), while Carno in both Argyll and Wales suggests a related Brittonic / Cumbric word.

Càrnach = a rocky place, as in Carnoch or Carnock and possibly Garnock, a cognate (derived from the same source) of the Welsh *carn* / *garn*, a prominence.

Carraig = a rock, cliff, pinnacle, promontory or headland (Brittonic and Welsh *carreg*), as in Carrick, Ayrshire, Carnoustie in Angus or Carrickfergus in Ireland, but sometimes it appears in the shortened old Gaelic *carr*, as in Carradale and maybe Crarae in Kintyre, or Crail, in Fife. There is also maybe a link here to an older word *kars*, rough, from an old Indo-European root *kar*, hard or stony, a word that is probably the root of the German word *karst*, a barren landscape shaped by the erosion of soft rocks like limestone. See **creag** below.

Ceann = head (pron. keown), the opposite of bonn / bun, usually written as **ken, kin,** or even **king** and replaces Brittonic *penn*, but it also appears in a few **kem** places, e.g. Kemnay, Aberdeenshire, from *a' mhaigh*, of the plain, or Kemback, Fife, from *bac*, a hollow. A related word, **ceap**, is a sod, stump, block, top or head as in Caputh, Inverkip or Keppoch and we have too many heads to count, but here is a list off the top of the head:

Kenmore = big head, as in headland
Kennoway = head, main field, from *ceann* plus *achaidh* (pron. achy), of a field
Kentra = head of the beach, from *tràigh*, beach
Kincraig = the head of the rock or crag
Kinghorn = headland of the mud or marsh, from old Gaelic *gronn*, marshland; Kingussie = (pron. Kinyoosie) the head of the pine wood, from *giùthsach*, pine wood
Kinlochleven = head of Loch Leven, probably from Gaelic *leamhan*, elm
Kinnaird = head of the point or promontory, from *àrd*, height or promontory
Kinross = head of the cape or promontory, from *ros*, cape or point

Kintail = head of the salt water, from *an t-saille* (pron. an tal-uh), of the salt water

Kintore = hillhead or head of the hill, from *tòrr*, steep hill

Kintyre = headland or end of the land, from *tìr*, land

King Edward (Aberdeenshire) was not named after an English king, but is an anglicised version of *cinn eadaradh*, meaning at the head of the division; Kinglassie, in Fife, was not the king's daughter, but a place at the head of a stream, from Brittonic *glas*; and Kingskettle was not what he used to boil water, but probably a high place belonging to a Briton called Catel / Ketil, or a place he went for peace and quiet, from Brittonic *cuddial*, a place of retreat.

Clais, claise (gen., pron. clash) = a ditch, trench or furrow, as in Cleish, Fife, and we need plenty of these in Scotland, though *cos, coise* (gen.) is a hollow or cavern, as in Gartcosh, an enclosure hollow. **Bac** places are also hollows, pits, bogs, peat banks, as in Bac Mòr or Back on Lewis, and we do not lack for **lag** or **glac** places, hollows or small valleys, as in Lagg, Arran, Laggan (a little hollow), or Logie, a hollow place (from Gaelic *lagaigh*). **Toll / tuill** (pron. towl) is a hole or hollow (but *tuil* means a flood), as in Clachtoll, a stony hollow, while **brothag** is yet another hollow or ditch, as in Brodie.

Claon (pron. klun) is a slope, as in Clyne, Clynder, Cluaine, and Claonaig, a little slope, or slope of the bay, and the old Gaelic *faithir / fothair* is a terraced slope, as in various north-east fetters places, e.g. Fettercairn, a wood slope (see Pictish **cardden**), Fetteresso, a soggy or waterfall slope (Gaelic **easach**) or Fetterletter, a tautological slope on the slope, Dunottar, a fort on the slope, but Forfar was an outlook or watch hill slope, from *fàire*, watch or sentinel. See also **leitir** and **sliabh** below.

Cnoc (pron. kroc or knoc) = a round hill, often supplanted by the anglicised **knock** in the Lowlands and north east, even appearing as Big Knock, Low Knock or just Knock, names applied by folk who no longer understood Gaelic, and likewise in Knockhill, i.e. a hill hill, an example of a pleonastic or knock knock name! We have about three thousand cnocs / knocks in Scotland, many qualified or described by colour, such as Cnoc Donn or Dubh, brown and black hill, or by size, shape, position, function or type of flora and fauna on them, as in Cnoc nan Gobhar / Gabhar, hill of the goats, in many places, or Knockando, market hill, from *ceannachd*, buying, as in the older version, Knockandoch or Knockander, though maybe from *Cnocan Dubh*, little black hill, in contrast to Greenock, a sunny hill, from *grian*, the sun, so no wonder so many cruise liners dock there today! Another wee hill, knob, lump or button is a **cnap**, as in Knapdale.

Coille (pron. CAWLyuh), cognate of the Welsh *celli / gelli* = a wood, forest, grove, as in Kelty (from *coilltean*, woods) and Kinkell in Fife (head of the wood) and possibly Coylton and Kyle in Ayrshire, but there are plenty more, often hidden in the woods, both as prefixes and suffixes. See also **kil** names. **Doire** (pron. DAWruh) is also a clump, grove, copse (often oaks), as in Northern Ireland's Derry, or possibly Durisdeer in Dumfries and Galloway, from *doire* and *ris* and *dubh*, *ris* being an older Gaelic version of *ros*, a wood, thus a copse of the dark wood. See also **bad**. **Craobh** is the Gaelic word for a tree (pron. croeuv), e.g. Crieff or Auchencruive, field of the tree.

Corrie (from Gaelic *coire*) = cauldron, kettle or pot, usually a mountain hollow, and **coirean** is a little corrie. Thus we have Corrie in Arran or Corrie Cas, a steep corrie, Corrie na Cloiche, corrie of the stone, or Corrie na Ciste, corrie of the chest, kist or coffin, Corgarff, the rough corrie, from *garbh*, Corrie an t-Sneachda, from *sneachd(a)*, of the snow (pron. shnechc / shnachkuh), or Corrour, the dun-coloured corrie, from *odhar*, brown. The Corrieyairick Pass is the pass of the rising corrie, from *èirich*, rising, climbing, but Corrievreckan is a whirlpool, the pool of Brychan (as in Brechin), a legendary warrior who challenged this pool and lost! Another nearby whirlpool is at Connel, from *coingheal*, referring to the tidal rapids and whirlpools at the entrance to Loch Etive, just below the bridge. Not a place to go swimming!

Creag, crag, craig = rock, cliff (*craig* or *graig* in Welsh) and there are millions, often named according to shape, size, colour, or after birds and animals, as in Craigellachie (the war cry of the Clan Grant), rock of the stony place, from *eileachaidh*, stony (a place-name also found in Canada), or Craigmillar in Edinburgh, not the rock of the miller as it might seem, but rock of the bare height, from *maol*, bare, and *àrd*, height. See **carraig** above.

Cuach (pron. kuach), gen. *cuaiche* (pron. quaich) = a cup, as in a quaich for drinking out of, but it can also mean a hollow place in older Gaelic (see **clais** above), as in Loch Quoich (of a hollow) in Knoydart in the west Highlands, Dun na Cuaiche, Inveraray and maybe Quoig in Perthshire or Cochno, near Duntocher, a place of cups, i.e. as on a cup and ring stone.

Cùil(e) (pron. coolu) = nook, corner, secluded spot, as in Culzean, nook of the birds (from *eun* / *eòin*, of the birds) or in its plural form *cùiltean*, possibly as in Cults places and **cùl**, meaning back place or hill back, as in Culbin (Moray) and Cùl nan Cnoc (Skye), both hill back places, while Coulport is maybe a backport (or from *caol* as in narrows) and Culloden, possibly means back of the little pool, from *lodair*, a little pool. *Cùl* can be disguised in Maryculter and Peterculter in Aberdeenshire, meaning back lands of Mary and Peter (from *cùl* and *tìr*, land), though easier to spot in

Coulter. Confusions also sometimes arise with *coille*, wood or *coll* or *calltain*, hazels. The Scots equivalent, almost a direct translation of *cùl tìr*, is Glasgow's Hyndland, a back land, though people there today are far from backward!

Druim (pron. DROYeem and often appears as **drum** or **dron**) = ridge, hump or back, as in Drumadoon, Arran, ridge of the fort, Drumbeg, little ridge (*beag*, small), Drumbuie, yellow ridge (*buidhe*, yellow), Drumchapel, ridge of the mare's back, from *capall*, mare or horse; Drumochter, upper ridge from *uachdar*, upper, or Belladrum, ridge at the mouth of the ford, from *beul àtha druim*. It also sometimes appears in disguise as Dron and Drem, or *dronnan / dronnaig* names, like Drungan or Drongan, little humps, so a humpy, but hopefully not a grumpy lot.

Eas, easa (gen.) = waterfall, cascade, ravine, as in Eas Fors Waterfall on Mull, all three words meaning the same (*fors* = a waterfall in Norse), or Bunessan on the same island, from *bonn*, foot, and *easain*, a little waterfall, or Eas a' Chuall Aluinn in Assynt, Britain's highest waterfall (from *cuil*, neuk, and *alainn*, beautiful) or Polnessan, in South Ayrshire, from B / G *poll*, a pool or hollow, and G *an easain*, of the little waterfall, from the diminutive *an* ending.

Eilean, eileanan (pron. EHlan) = island(s), the more modern Gaelic word, as opposed to the older word *innis* (see below) and is used mainly in the Western Isles and north west, such as Eilean Beag, wee island, Eilean Garbh, rough island, Eilean nan Gobhar, goat island, Eilean nan Ròn, seal island, or the famous Eilean Donan Castle (Donnan's Isle).

Gleann (pron. glown), anglicised to **glen** (*glyn* in Welsh) = usually narrower and steeper than straths, often named after its size, shape, the river that runs through it or a nearby mountain. Glen More is the Great Glen or Glen Albyn, the longest glen in Scotland, from *Alba / Albainn*, Scotland, of Scotland. Our landscape is covered in glens and nearly every town or village in Scotland must have at least one, or even several. A **sloc** or **slochd** is a deep hollow, pit or ravine, as in Slochd Mor, big ravine, but a **glac** is a hollow or small valley, a bit like a *bac* or a *lag*. See above.

Inbhir (pron. EEnvur / EEnyir) = river mouth, confluence, usually written as **inver** and replaces Brittonic Aber. There are over four hundred Invers in Scotland, e.g. Innerleithen, confluence of the Leithen with the Tweed (Inver has been anglicised to Inner). **Comar** is another confluence word (a cognate of Welsh *cymr*), as in Comers, Aberdeenshire, Comrie, near Crieff, or Cumbernauld, from *cumar nan allt*, joining of streams, and another one is to be found at Condorrat, in North Lanarkshire, from *comh*, joining, and *dobhar*, water, and *àite*, a place. Inveraray is the mouth of the River Aora (pron. ayra), from an ancient word Aray, Aar or Ara, possibly meaning flowing or smooth-running water (see Pre-Celtic river names). Inveresk is the mouth of the Esk, a pre-Celtic or early Celtic river name *esc*, as in Gaelic *uisge*, water. Inverkeithing is the mouth of the Keithing burn, from Brittonic *coed*, wood, and Inverness is the mouth of the River Ness, another pre-Celtic name that predates the Norse *ness*. Inverurie is the confluence of the rivers Don and Urie, the latter possibly from Gaelic *iubharach*, of yews or *uar*, a landslip or spout, or from *uidhre*, meaning darker, browner. Sometimes *inbhir* appears at the end of a name as the qualifying element, e.g. Lochinver, loch mouth, or Kilninver, church at the confluence, from *cill an inbhir*. **Ùidh** is another river mouth word, a Gaelic version of a Norse word which can also mean an isthmus or stream joining two lochs, as in Loch na Garbh Ùidhe in Assynt, a loch with a rough burn between two lochs.

Innis, innse (gen.) (pron. EENish), *ynys* in Welsh and *enys* in Cornish, often shortened or Scotticised to **inch, insch,**

insh = an older word for island (see **eilean** above) which now usually means a water meadow or grassy island beside a river, or a small island, often in a loch, e.g. in Loch Lomond we have Inch Cailleach (Island of old women / nuns, from *cailleach*), Inchmurrin or Innis Mearain (Mirren's island), Inchcolm or Innis Colm (St Columba's island) in the Firth of Forth, or Markinch in Fife, meaning horse island, from old Gaelic *marc*, horse, while we have Inchnadamph in the Highlands, from *innis nan damh*, island or meadow of the stags or oxen. There is also Insch in Aberdeenshire, or the broad river meadows of the Inches in Perth, beside the Tay, and further up the A9 road there is a place called Phones, an anglified version of *fo-innis*, meaning an under or smaller meadow / field. The suffix *isidh*, meaning a meadow, is derived from *innse*, of a meadow, as in the Glen and River Feshie.

Làirig (pron. laarig) = a pass, beaten path, usually between hills, as in Lairg, Lairig Ghru in the Cairngorms and maybe Crianlarich, but **learg** is a sloping plain or hillside, as in Largs, or Largo, from *leargach*, steep, but **làrach** is a house site, maybe a ruin. Arran has three largys on the south-east of the island: Largy More, Largy Meanoch and Largy Beag, a big, middle and small one, probably all from *learg*.

Leitir (pron. LEHtchir), from *leth tir* = half land, slope, hillside (Cumbric *llethr*) as in Letters and other letters places which have nothing whatsoever to do with reading and writing, e.g. Letterfearn, slope of the alder trees or Letterewe, slope of the River Iu, old Gaelic for yew tree. Both **leachd** and **leathad** refer to a declivity, slope, pass or hillside, as in The Lecht, Ledmore and Ledbeg, big and wee slope and possibly Ben Ledi, though also thought to be from *le Dia*, with God! See also **ruighe / ruidh**.

Linne (pron. leenyuh) = a pool, channel, sound or bay, as in Loch Linnhe, just meaning the pool, Cora Linn, or Falls of

Clyde, from Gaelic *corrach*, marshy, boggy. Lin or Lynn is more common in the Lowlands, as in Lynn Glen, though some 'lins' refer to lint or flax, as in Linton. See Brittonic **llyn**. **Lòn** can also mean a pool (as well as a brook, marsh, meadow, a blackbird and other things), while **lèana** and **lèanag** are also meadows, though swampy or small ones, as in Leana / Lianag Mhor, both bigger varieties of such places.

Loch – Scotland has 31,460 lochs, plus one 'lake', the Lake of Menteith, though this 'lake' is a corruption of the Gaelic word *leachd*, sloping ground, replaced by the Scots word *laigh*, low. Undoubtedly, Loch Fyne is the finest, as it is possibly a loch of wine, from old Gaelic *fine*, wine (modern Gaelic *fion*), but, like the glen and river, it could also be a virtuous place, a name derived from an ancient holy site, though other meanings are possible and many have doubts about wine and virtue being combined! What seems like a tautology or double loch, Loch Lochy, is apparently the stream of the black goddess (see pre-Celtic river names). In contrast, Lochgelly is a bright shining loch from *geal*, bright and Lochnagar is from *na gàire*, loch of the noise or laughter, and no wonder it is laughing, with a mountain named after it which clearly isn't a loch! Lochinver, at the river mouth (*inbhir*), needs to be distinguished from Lochinvar, which looks similar, but is from *an bharra*, of the height, where Sir Walter Scott's hero young Lochinvar resided. Lochmaddy means loch of the wolves or dogs (*nam madadh*) and thankfully not the loch of the madman!

Machair = sandy plain, grassland beside the sea, e.g. Machrie, Arran or Machrihanish in Kintyre, and the Western Isles are covered in them.

Maol (pron. moeul) = bare top, headland, round hill, point or promontory, literally a bald head or brow of rock (also Cumbric / Welsh *moel* or *foel* as in Melrose) which features in many hill or headland names, such as Maol Buidhe, a yellow brow in Ardnamurchan, or Milleur Point, Wigtonshire, a brown promontory, from *odhar*, dun coloured, or the Mull of Galloway or Mull of Kintyre, while the Ross of Mull is a tautologous headland (see Brittonic **ros**). Maol is sometimes anglicised in the Lowlands as *mill* and, just as some men try to disguise their hair loss, it is well disguised on the Oa in Islay, from *maol* and Old Norse *haugr*, a rock or cairn and, probably because that is a bit of a mouthful, it was eventually contracted to *oa*, meaning a headland or brow of rock. However, some mull places, especially in Orkney and Shetland, are from the Norse word *muli*, a headland. **Meall** (pron. meeowl) or **mill** in its genitive or plural form, is a similar-sounding word with a similar meaning of lump, heap, mound or round hill and we have heaps of them – while *meallach* means lumpish and the similar **mullach** is a roof, top, summit or ridge as in Mullach Buidhe on Arran.

Mòine (pron. moniu), **mònach** (gen.) and also **mòinteach** = peat bog, moss, moor, as in Moniaive, possibly means moor of crying, from *èibhe*, cry, Montrose is the bog of the promontory from *ros*, Drumoyne, fort of the bog, and Monymusk is a foul bog, from *mosach*, filthy, foul and Scotland has no shortage of them. A related or overlapping word is **monadh** (pron. MAWnugh = hill or hill pasture, moor. Our **mon** / **mona** names do not come from French or Italian, but from monadh, as in Monamore, a big hill on Arran, Moncrieff, wooded hill, from *craobh*, a tree, Minnigaff, hill of the smith from *a' ghobhainn*, or Monadhliath, grey mountain(s), Am Monadh Ruadh, the red mountain (or Cairngorms), or Crimond, from old Gaelic *crech*, a hilltop, so it is a really top mountain. The Welsh cognates are *mawn*, *mynydd* or *fynydd* and some of these names may well go back to a Brittonic source.

Poll / puill (gen.) = pool or pit (see also Brittonic) or mud and mire, though older *pol* names have the meaning of flowing water. Pollok in Glasgow is just a little pool, Pollokshaws is a little pool in the woods, while the nearby Polmadie is maybe the pool of the dogs, from *madadh / madaidh*, hound(s), of dogs, as in Polmaddie Hill in Ayrshire, though it is also claimed to be from *mac Dè*, son of God, a name which illustrates how the original Gaelic stress pattern has survived for centuries in the tongues of non-Gaelic speakers, as it is still called 'PawmaDEE', with a half stress on 'paw' and a full stress on 'dee', just as it would have been said in Gaelic. *Poll mònach* means a peat bog and some places are up to their necks in them. See also **llyn** and **linne** and **lòn**, for more pools.

Rèidh(e) (pron. ray) refers to a flat or smooth piece of ground, whereas **ruigh(e)** (pron. rooyu) means a forearm, lower slope, grazing or run for cattle, sheiling, e.g. possibly as in Rhiconich, from *chòinnich*, moss (though it is possibly from *rubha*, below), or in Portree, the port of the slope, or Dalry (see **dail**), and its river Rye. See also **leitir**.

Ros, rois (gen.) = headland, wooded promontory, peninsula, as in Br *ros* where it can also be a wood or moor. In Arran, the Ross and the hill road of that name refer to the moorland hill or headland above Lamlash, while Ardrossan, the main ferry port for Arran, means the little high cape or headland, from either the Br *ros* or the Gaelic word of the same meaning.

Rubha, rudha (pron. roouh) = a spit, headland or promontory, a very common coastal feature on the west coast and the islands (including Arran) which have rubha names galore as in the anglicised Rhu in Argyll or Rhu Bodach (old man) on Bute, Rubha Mòr on Barra, Rubha na Creige Mòire (headland of the big rock or cliff) on South Uist or Rubha Clachan (of the stones) on Kintyre and, since they are such rocky places, many others use *clach* or *leac* (flat stones), such as Rubha Leacach on Harris or Rubh' an Dùnain (of the wee fort) on Skye and Lewis. They are quite often related to fauna, such as Rubha nan Cearc (hens, chickens) on Mull and Skye, Rubha na h-Easgainne (point of the eel) on Skye, Rubha na Ròinne (of the seal) on Rum or Rubha nan Sgarbh (of the cormorants) in Kintyre. See also Brittonic **ros**, Norse **nes** and Gaelic **sròn**. Many Rubha names have been changed to Point names by mapmakers.

Sàil, sàl, sàile, sàlach (pron. saahl) = a heel, as in various Salen places, or Sàil Garbh, a rough heel, while *sàilean* (pron. sahlen), *sàilein* (gen.), is a little inlet, arm of the sea, a bay, as in Bàgh an t-Sàilein, bay of the little inlet, or just a bay bay!

Sìthean / Sìdhean (pron. sheehan), appears as shee / schie = a conical hill, hillock or fairy knoll, as in Glenshee, or from *sìthe*, of peace, reconciliation, or sometimes from *sitheann*, venison, game. Schiehallion possibly means the fairy hill of the Caledonians, or alternatively maiden's pap, from *sine*, breast and *cailin*, of a girl. Sìthean Beag is a little hillock and Sìthean Mòr a big one and we still have plenty of little people lurking in our landscape. **Digh** (pron. ji) is another word for a conical hill, rampart or abode of fairies.

Sliabh (pron. shleeve), **slèibhe** (gen.) = mountain, hill, slope, moor, a very common word in Ireland, where it usually means mountain, but less so in Scotland, though it arrived with the earliest Scots settlers and took root in various forms in the southern Inner Hebrides, Kintyre and Arran, and in a few other places, such as Buchlyvie in Stirlingshire, meaning hut on the slope (from *both* and *slèibhe*), or Slamannnan, hill of Mannon or Manau (as in Isle of Man), a Cumbric name for the area around the River Forth. We have Sliabh Fada, long hill, Sliabh Meadhonach, middle or central hill, Sliabh Mòr, big hill, Sliabh nan Dearc, berry hill and Sliabh na Mòine, peat hill. It also appears in **slew** hill names in western Galloway (where an early Scots colony existed before Norse Gaels arrived), sometimes perhaps meaning a moor, as in Slewbog, but more often just a hill as in Slewmuck, hill of the pigs, from *muc*, another mucky place.

Srath(a) (pron. srah), usually appearing as **strath** = a flat, wide valley, as in Strathdon, Strathclyde, Strathspey, Strathglas, a stream vale or valley burn (*glas* meant water in an older Brittonic form, but in Gaelic, just to confuse things, it is grey or green), Strathaven (pron. Straven), valley of the River Avon, Strathpeffer, strath of the shining stream, from Cumbric *pever* (as in Peffermill, Edinburgh).

Sròn (pron. srawn), **sròine** (gen.) = nose, point, as in Troon, from Brittonic **trwn**, meaning the same, or Stranraer, a thick nose, from *reamhar*, thick. It is often anglicised to **stron** as in Strone, Argyll, Sròn na h-Iolaire, eagle point on Rum, Stronlarig, Stirlingshire, from *sròn làirig*, point of the pass and maybe Strontian, from *sron teine*, promontory of the fire or beacon and from which the mineral strontian was named as it was first discovered here, though the name could also be from *sròn an t-sìthein*, the nose of the fairy hill! **Soc**, a snout, is a related word in meaning, from which we get Succoth, a point of land at which two streams meet and so it sticks out like a nose. **Corran** is another point word, a sickle, crescent or point of land running into the sea or to an island, visible only at low tide, as in Corran Ferry. **Rinn** or **ruinn** (see Cumbric **rhyn**, pron. ryne) is also a point, though more of a tail, as in Rhynd, Perthshire and also maybe the Rinns of Galloway and Islay. See **roinn / rann**.

Sruth, sruthan (pl) (pron. stroo(an)) = a stream or burn (and *bùrn*, pron. boorn, is an alternative Gaelic word for water, see **uisge**), anglicised to stru, meaning current or stream, maybe as in Stirling, meaning enclosure by the stream, from *lann*, enclosure, but certainly as in Struy, Struan or Anstruther in Fife, from *an sruthair*, the little stream, known locally as Ainster / Enster.

Tarbert, tarbet, tarbat, from *tairbeart* = a narrow neck or isthmus over which boats and goods could be rolled or dragged to the next loch or the sea, as in many Tarbert places or Loch Tarberts, as in East and West Loch Tarbert in a number of places.

Teanga = a tongue or language, but can be used for a spit of land, as in some Changue places. See ON **thang, tunga / tangi**.

Tìr, tìre (gen.) (pron. cheer) = land (*tir* or *dir* in Welsh) as in Kintyre, meaning head or end of the land, Tiree, possibly meaning land of corn (from old Gaelic / Irish *iodh* or *eadha*, of the corn), though much debated and maybe from an old Irish name *Ith*, meaning land of Ith. Tirafuir, probably means cold land (from *fuar*) but Tìr nan Òg is the land of the ever young (i.e. Paradise).

Tòrr (pron. tawr) = large heap, mound, rock, conical shaped hill, great amount of anything, as in Torness, East Lothian (from Norse *nes*, headland) or Tornapress, Highland (from *na preas*, of the wood), or various places in Aberdeenshire, such as Torphin(s), a white hill, (from *fionn*, white or fair), Torry, a hilly place, with the suffix *aidh*, meaning of a place. Tòrr appears in the name of one of Scotland's most famous literary characters, Burns' Tam O Shanter, as Shanter, the name of his farm, is probably from *seann tòrr*, or old mound. **Torran** is a little hill and we also have piles of them.

Tràigh (pron. try) or the genitive **tràighe / tràghad** (pron. trahad) = a sandy beach, seashore, lochshore, e.g. An Tràigh

Mhòr, the big beach, Barra's cockle strand. **Gaineamh** (pron. *gan*yuv) = sand, as in Ganavan Bay, near Oban, while **muir / mara** = the sea / of the sea and **cuan** is the ocean or sea, as in Cuan nan Orc (ocean of the whales), i.e. The Minch. **Cladach** is a stony beach, as on Cladach, Arran and many other places, and **mol** is a shingly beach, while **bàgh / bhàigh** is a bay (see **bàgh** above) and **sàilean / sàilein** is a deep bay or an inlet / arm of the sea (see **sail** above). **Dòirlinn**, another beach word, refers to a tidal isthmus or tombolo, as in An Dòirlinn on Lismore, or at Castle Tioram in Moidart, or the island of Earraid (Ross of Mull), where Robert Louis Stevenson's hero, David Balfour, was temporarily marooned in *Kidnapped*. The Norse word is ayre / ire. **Oitir** (pron. oytchir) is a tidal sandbank, ridge, reef or low promontory, as in Otter Ferrry on the Cowal Peninsula or Oitir Mhòr, the big sandbank on the island of Kerrera, near Oban.

Tulach or **tulloch** (pron. tooluch) = knoll, hillock, mount, as in various **tull** or **tully** names, as in Tulloch, Tullibody, from *a' bhothag*, hill of the hut or bothy, Tullybelton hill of Bealtainn or Beltane, the May fire festival, or Tullochgorum, greenish hill, from *gorm*, blue-green, and sometimes appears as **tilly** as in Tillycoultry, hill of the back-land or settlement, from *cùl* and *tìr* (or maybe Brittonic *tref*) and in Tillienaught, a bare hill in Aberdeenshire, from *nochd*, bare.

Uamh (pron. uahv), **uaimh(e)**, **uamha** = cave, as in Uamh Fhliuch, a wet cave, Loch nan Uamh, loch of the caves, a place used by Prince Charlie in 1745–46, and via some sound changes it transmuted into Wemyss Bay or Port Wemyss.

Uisg(e) (pron. ooshcu) = water (also **bùrn, bùirn** (gen.) for fresh water) as in Loch Coruisk on Skye, meaning hollow or corrie of water and various Esk names, some possibly related to *easg*, a marsh or swamp.

And Many More Mountain Words

Many Gaelic landscape words appear in both their Gaelic and their anglicised forms, giving us pairs like bàrr and bar, beinn and ben, cnoc and knock, druim and drum. Most of these can be checked online, e.g. the Ordinance Survey or Scottish Place-names Society / Scottish Language Dictionaries websites, especially the many words for particular types or shapes of mountains, such as *biod*, *biodan*, pointed top(s), as in Beattock, *bioran*, a little pointed stick, sharp pointed peak, *cìr*, a comb, crest, as in Cìr Mhòr, Arran, *faire*, a summit, horizon, lookout, or watching, as in Cnoc na Faire, *stac*, a steep conical hill or peak; *stob*, a point, post, stake, stump, *stor*, a peak, post, pinnacle, *stùc*, a pinnacle, peak, steep rock, a *sgòrr* or *sgùrr*, a rocky peak, found a lot on Arran or Skye; *òrd*, a hammer or chunk, as in Beinn nan Òrd, or *guala(i)nn*, a shoulder, long ridge, as in Gullane; *tom*, a round hillock, round heap, as in Tomintoul, a barn-like knoll, *cruach*, a heap, stack or haunch, as in Ben Cruachan or *tòn*, *tònan*, bottom(s), backside(s) or *cìoch*, a breast, as in Cìoch na h-Òighe, or Cìoch na Maighdin, both meaning breast of the maiden, while we also have *màm*, a breast shaped or large round hill, as in Mamore and Mambeg, the large and small variety!

Settlement Names

Achadh (pron. ahchugh) = a field, but eventually implying a farm settlement, usually written **ach, auch** or more often **auchan, auchen, auchin**, from *achadh an / na*, the field of, though it can also be from the diminutive ending *auchan*, a wee field (though the suffix **ach** or **och** refers to a place or a share, as in Coigach, a fifth share, from *coig*, five), while it also appears as a suffix in Breakachy, from *breac*, a speckled field or Keppoch, a top field, from *ceap*, top. Ach / auch names are very old as they go back to a time when a place was defined or located by the field or fields around it, though most of them eventually became settlement names. They are so common across the whole country (second most common, after *baile*), apart from the Outer Hebrides, that there must be thousands of them, but here are some examples:

Achiltibuie = field of the yellow-haired lad, from *gille*, boy, and *buidhe*, yellow
Achnaclach or Achnacloich = field of the stone or stony field. See **clach** below
Achnasheen = field of storms, from *sian*, storm
Auchanalt = field by the burn, from *allt*, burn, stream
Auchencairn = field of the cairn
Auchencruive = field of the tree, from *croibhe*, of a tree
Auchenheath = field of birch, from *beithe*, birch
Auchenleck = field of the flagstones, from *leac*, a flat stone, slab or grave-stone
Auchenshuggle = possibly field of the rye, from *seagail*, of rye
Auchentiber = field of the well, from *tobair*, of a well
Auchmore = big field, from *mòr*, big
Auchnashellach = field of the willows, from *seileach*, *seilich*, willow, of willows

See also Pictish **pit** names and Cumbric **llan** / Gaelic **lann**.

Àirigh or **àiridh**, anglicised to **ari** = sheiling, bothy, hill pasture, again very common, especially in the Highlands and Islands, where the work of moving and herding animals up on the summer pastures was often the cultural highlight of the year. Many are defined by location, as in Arichamish, from *Àirigh a' Chamais*, the sheiling by the bay, others referring to people, occupations, or activities, such as *Àirigh Mhuirich Cheàrdaich*, the sheiling of Murdo's smithy in Braeleny, Stirling, while many are related to flora, fauna or colour, such as Airth, a level green place or sheiling, Glassary in Argyll, a grey sheiling, from *glas*, grey, or Ardrishaig, *Àird Driseig*, the high briar or bramble sheiling by the bay, found in several places, from *àrd* / *àird*, a height, *dris*, a brier or bramble, plus Norse *vik*, a bay, or *Àirigh nan Cuileag*, sheiling of the flies, Perthshire, where the midgies probably drove them all daft!

Annaid, annaite (gen.) (pron. ANNetch), often **annat** = mother church or patron saint's church, or a church with relics, often with ancient links, possibly to pagan holy places associated with the goddess Anu, especially beside water, needed for baptism, as in Allt na h-Annat on Ben Dorain and Ben Lawers, or Achnahannet, church field, Annatland, Annathill, or Longannet, from Cumbric *llan* / Gaelic *lann*, a field, enclosure, and *annat*, church, or just plain Annat in Argyll.

Baile (pron. balla), our most common settlement name, usually written **bal, ball** or even **bel** = enclosure, farm, village, town and replaced Pictish *pit* or Brittonic *tref*. Bal or bally (very common in Irish place-names) is very old and crops up everywhere over a long period of time, as in Balado, long village, from Gaelic *fada*, long, Bailbeg, a wee farm, from Gaelic *beag*, little, Balfour, pasture village, from Brittonic *pawr*, pasture, Balgowan, Balgownie, Balnagowan, Balgonie, all villages of the smith(s), from *gobha*, *gobhainn*, blacksmith(s) (pron. go-u, gowan), likewise Balnaguard, village of the craftsman, from *ceàrd*, craftsman or black-smith, or Balintore, from *an todhair*, of the bleaching green; Ballantrae is a village on the shore, from *tràighe*, of / on the beach, Balerno, sloe tree or blackthorn farm, from Gaelic *àirneach*, Balmaha, village of St Maha, another early Celtic holy man; Balmoral is village on the big clearing, from Gaelic *mòr* and Brittonic-Pictish *ial*, a clearing,

Balquidder, fodder village, from old Gaelic *foider*, fodder, and finally, the famous Glasgow Bella who was not a woman, as Bellahouston was once a village of the *ceusadan*, cross or crucifixion.

Blàr, blàir (gen.) = cleared space, plain, battlefield, originally a clearing or space that was sometimes used to settle disputes, a prefix found in many names, e.g. Blair Castle, Blair Atholl (see **ath** above), Balblair, village of the plain, or Blairadam, not the location of the Garden of Eden, but, via several sound changes, plain of the oxen from *blàr nan damh*, or Blairgowrie, either named after Gabran, a sixth-century king or the district of Gowrie, one of the divisions of Pictland. **Cath** is another Gaelic word for a clearing, or a battle place, maybe as in Catrine, Ayrshire, though debateable.

Both, bothan (pron. bo and bohan) = turf house, hut, sheiling, bothy, shelter, and the similar **bùth, bùthan** is also a bothy, hut, cottage, or a booth, tent or shop (and on Lewis, in former times, an illicit drinking den), as in the many bothies on our mountains or in places like Bonhill, as it was a house by / of the stream, from *both an uillt*, or Buchlyvie, Stirlingshire, a hut on the slope from *slèibhe*, of / on a slope, or Bowmore in Islay, possibly just a big hut. Unfortunately, Bochastle, Stirling, is not quite as grand as it might seem, as it is just the castle hut, bothy or howf, from Gaelic *a' chaisteil*, of the castle, while Boclair, near Bearsden, Glasgow, is not quite as French as it sounds, though it may have been a cleric's hut (as with the name Buchanan, the Canon's hut), but maybe it was just a hut on a flat piece of ground (Gaelic *clàr*, board, table, level ground), or even the flat ground for cows, as *bò* words can refer to cows. *Both* is related to the Cumbric *bod*, Norse *bud* and Old English *botl* / *bothe* / *booth*, maybe as in Bothwell. This is where we get a *but an ben* from (old cottage with an inner living part and outer part for animals) where the Broons went their summer holidays!

Cair, caer, kir – see Brittonic section.

Ceapach = a plot of land or tillage plot, with the *ach* suffix meaning a field, as in Ceapach and Keppoch, from **ceap** (pron. kepp), sod, piece of turf, stump, block, head, cap, as in Barkip, Inverkip or Kippen, a little stump or block.

Cill / cille (gen.) (pron. keel, keeluh), usually written as **kil** = church, churchyard, originally the little cell / shelter (*ceall* in Gaelic from Latin *cella*) in which the monks or missionaries of the early Celtic church lived the simple life, and *cill* names emerge as soon as churches were established in Gaelic-speaking communities from the sixth and seventh centuries onwards. The second element often refers to the inhabitant of the cell or the saint to whom it was dedicated, but the early church often grafted saints' names on to older pagan ones (see below) in order to venerate their new saints and eradicate the older beliefs, though sometimes kil names only refer to a church place, as in Kellas (*ais*, a Pictish place). We have no shortage of them, e.g.

Kilbrandon = church of St Brendan or Brennan
Kilbride = church of St Bride or Bridget, a name derived from the old Celtic goddess Brigit
Kildonan = church of St Donan
Killean = Church of St John, from Gaelic *Iain*, John
Kilmacolm = church of my Colm, i.e. church of St Columba
Kilmarnock = church of my little Ernan or Ernoc, i.e. St Ernoc, from *mo*, my, and Ernoc
Kilmartin = church of St Martin
Kilmory = church of Mary, from *Moire* (i.e. the Virgin Mary)

Kilvaxter (Skye) = from Cill a' Bhacastair, as in Scots *baxter*, the baker's church, maybe a baker who built or baked in a church or a congregation famous for its baking!

Sometimes the second element describes the location, e.g. Kilcreggan meaning church on the little crag or rock or Killin, white church, from *fhionn* (fh is silent), white. N.B. **Cill** is often replaced by **kirk** (see Norse and Anglic / Scots sections), and sometimes **kell** or **kill**, as in Kelburn, or Kill Brae, while conversely **kil** sometimes replaces or hides other Gaelic names, such as Kilbrannan which is actually from Caol Brenaind, Brandon's kyle or narrows, though Kilbrandon on Seil Island is Brandon's church, as Kilbrennan on Mull probably is. Kilcoy is from Cùl Coille, back of the wood, and most Killie names come from *coille* (wood) as in Killiecrankie, aspen wood, from *coille creitheannich* (aspen) and also possibly Kyle in Ayrshire. The other Gaelic word for a church is *eaglais* – see Cumbric.

Clach, cloiche = stone, of stone, e.g. Clachan, stone house or village and also as in Clachnacuddin (stone of the tub, *na cùdainn*), Clackmannan, stone of Mannan, possibly an ancient divinity or a personal name, Achnacloich, stony field, Clauchlands and other clauch places on Arran, Cloich Hills (Borders) or Stronachlachar, Loch Katrine, the mason's nose or point, from *sròn a' chlachair*, but there are clach places wherever you look.

Cladh (pron. clugh) = a churchyard, cemetery, trench, as in Cladh nan Sasunnach or English cemetery and **ciste** is a coffin or chest, as in Bealach na Ciste, coffin pass, though Kiscadale, coffin glen or dale on Arran, is from the Norse *kistu* and *dalr*. It's really a dead end word.

Crìoch (pron. CREEuch) = a boundary, border, march, end or limit, as in several places called Creich, though some could be tree places from *craobh*, a tree or *critheach*, aspen. It also features in Cree, the river and the place that takes its name from it, Creetown, similar to Crichton, a boundary farm in Midlothian, or as in Sròn na Crìche, the boundary promontory, on Loch Tay.

Dail, dal (pron. dahl) = field, meadow, haugh, fertile land often beside a river, so a water meadow, but also implying a settlement, a word common to Cumbric (*dol*) and old Gaelic (*dal*, which could also mean a portion, share or tribe, as in Dal Riada), Norse (*dalr*), English (*dale*). If it's from Gaelic, normally it's used as a prefix (before the name), but if used as a suffix (after the name) usually it's Norse or English, as in Helmsdale, Clydesdale and it usually means a dale or valley, though we find places like Knockandale, near Symington in Ayrshire, where it appears as a qualifying word, a name which has preserved the Gaelic grammatical structure in spite of anglifying the *dail*. See Norse **dalr** and Anglic **dale**.

You can't avoid them because they are all over the countryside, such as at Dalbeattie, field of the birch, from *beithe*, Dallas, from *eas*, waterfall, or Pictish-Brittonic *ais*, a place (a name that travelled all the way to Texas), Dalmillin, field of the mill, from *muileann*, Dalmuir, big field, from *mòr*, big, but confused with Scots muir; Dalrymple, meadow of the crooked or winding pool, from *crom*, bent, curved, and *poll*, pool, Dal na Spidal, field of the hospital or refuge, from *na spideil*, Dalziel (pron. Dalyell), white field, from *dal* and *geal*, white.

The name Dalry is found in several places and is often thought to mean king's or heather meadow, from *righ*, king, or from *fraoich / fhraoich* (fh is silent), heath, heather, of heather, which is what it says on the railway station in Dalry, North Ayrshire, but it is far more likely to be from *ruighe* (rooee-yuh) meaning a slope, so it is probably a sloping field, meadow or haugh, though *rèidhe*, level, smooth, has also been suggested. **Lèana** and **lèanag** are also meadows or leas, though swampy or small ones, or even lawns, as in Lèana / Lianag Mhòr, both bigger varieties, while *àilean* is another meadow or enclosure word, e.g. Glac an Ailein, a meadow hollow on Mull.

Davoc, davoch (from Gaelic *dabhach*) = tub, vat, tract of land, a land measure and usually appears as **doch, dauch**, sometimes shortened to *och* and used in former Pictish areas, e.g. Dochgarroch, rough land, from *garbhach*, rough, Dochfour, pasture tract, from old Gaelic *phùir*, pasture, or Findoch, white land, from *fionn*, white, fair, bright. Haddo in Aberdeenshire is just half a davoch, a lesser place, though Haddo House certainly isn't! A quarter of a *dabhach* was a *ceathramh*, a fourth part, as in Kirriemuir, the big quarter (famous for its ball), from *mòr*, big, so maybe shares were not always equitable.

Dùn (pron. doon), **dùin** (gen.) = hill, mound, fort (as in Brittonic *din*), is one of the oldest Gaelic settlement words, some going back to the Br din / dun, sometimes changed to *ton* in the south east or appears as **dum**, as in Dumbarton, fort of the Britons, while Dunkeld is thought to be the fort of the Caledonians, both clearly named after their occupiers, possibly like Dundee, hill of Dèagh or Daig (a personal name) or from Dè (genitive of *Dia*, God), though maybe just from *dubh*, meaning dark hill. Sometimes a dun just refers to the size or shape rather than an actual fort, as in Dunmore, big hill, or the diminutive *dùnan*, a wee fort or a heap, including a heap of dung, while **dìong** (pron. jeeng) is also a wee fort, dwelling or hillock, but Dungeon in Galloway isn't a prison, but a fort or hill of shelter, from G *dion*, shelter, or defence.

Dùn is often followed by a word for a landscape feature, as in Dumfries, hill of the copse, from *phris*, genitive of *preas*, a wood, or Dunlop, hill of the bend, from *lùib*, of a bend, or Dunbar, fort on the height, from *bàrr*, height or top. Dunottar means fort on the terraced slope, from *faithir*, shelved or terraced; Dunoon is a river fort, from *abhainn* and Duntocher is from an old Gaelic word *tochar*, road or causeway; Dumfermline is hillock fort, possibly from *meallain*, hillock, or from a personal name, Parlane or Farlane. Sanquar (*sean caer*) in Dumfriesshire and Shandon (*sean dùn*) in Argyll and Bute are both old forts. An even older fort is Fortingall, with its three-thousand-year-old ewe tree, from old Gaelic *fartair* (from Cumbric *gwerthyr*), a fort, plus *cil*, a church. **Dùnadh** can mean a fort or a camp, as in Donibristle, Fife, the fort of a man called Breasail.

Earrann (pron. err-ann) = share, portion, area, often appears as **arn** or **iron**, as in Arnprior, Arnvicar, Irongath / Airngath Hills, a windy area, from *gaoithe*, of the wind, Irongray (from *greigh*, of the herd or stud), Killearn, apparently church share, though it was originally *cinn* and *earn*, meaning head portion. Similar sounding words are **fearann**, fertile land, farm or estate, as in Ferintosh, the chief's land (from Gaelic *tòiseach*, chief) and, in the Lowlands, *erne*, Old English for a house. See also **roinn**, another share or allotment word.

Fas or **fasadh**, sometimes shortened to *ais* = a cattle stance or level place, where drovers rested the cattle overnight, as in Faslane, an enclosure stance, from *lann*, or Faskally, stance by the ferry, from *calaidh*, Fasnakyle, from *na coille*, so stance of the wood, or Duffus, black stance, from *dubh*, or Barassie, a top of the level stance, from *bàrr*, top or crest, plus *fhasaidh*, of the stance. But **fasgadh**, a shelter or cattle fold, is not to be confused with **fàsach**, a wilderness, desert or stubble.

Gàrradh (pron. gaahrugh / garru) = garden, enclosure, dyke, yard or place, while in the Outer Hebrides the Norse *gardr* took root, instead of *achadh*, a field. We thus find it in various shapes and forms, such as Garscube, corn enclosure, from *sguab*, sheaf, or Gargunnock, enclosure of the rounded hill, from *cnuic* or *dùin-ock*. Similar words, or variants, are **geàrraidh**, green pasture land, common grazing, or the older Gaelic **gairt / gart / gort**, a field, cornfield or enclosure, as in Garth, a corn field, Gartocharn, enclosure of the humped hill, from *chàirn*, Gartmore, a big field, or Gartsherrie, field of the colts, from *searraich*, or Gartnavel, from *gart an abhaill* (of apple trees), an orchard, or Garscadden, herringyard, from *gart* and *sgadain*, herring, while **goirtean** is a small patch of enclosed arable land or park. The Cumbric / Welsh word is *gardd / garth* (as in Garth above), the Norse equivalent is **gardr / garðr**, while the Old English word is *geard*, as in Scots *gairden*.

Lann = an enclosure, church – see Brittonic section.

Lech (see Br **llech**) = slate, slab, tombstone, grave, but can also be used in landscape names, just referring to flat stones. It features in Auchenleck, the field of the slabs, or Leac Gharbh, rough slabs near Sannox on Arran, or Lix places, e.g. the one in Angus, from the plural *lic*. Confusion can arise as *leacan* can mean wee slabs, but **leacann / leacainn** is a hill slope, which can also be a cheek or brow as in a number of places on Arran.

Lios, lis = garden, or enclosure (*llys* in Brittonic), e.g. Leys, Lismore, big garden, Kinloss, head garden and possibly Lesmahagow, garden of one Mahagow, though the gow suffix could be from Brittonic *cau*, a hollow, while **lus** is a plant, weed or herb, as in Luss, Glenluce and other Luce places.

Magh, maigh (gen.) = field or plain, maybe as in Mauchline, from Gaelic *linne* or Cumbric *llyn*, a pool, Mawcarse in

Perth, plain of the carse, i.e. flat land beside a river, Moffat, the long plain, from *magh* and *fada*, long. It often appears as **mauch, maich** (meaning a low lying, boggy plain in Scots) or **mag**, as in Magus Muir, Fife, which might disappoint some as the word has nothing to do with a wise man or sorcerer, but just a plain old field with a tail, from *gasg*, a tail, while it is even more hidden in Monzie (pron. monee), a cornfield in Perthshire, from *magh an eadha*. *Magh* could refer to landscape as well as settlement. See also Brittonic **maes**.

Ràth – see Brittonic section.

Roinn(e) (pron. royn), **roinnean** (pl) and also **rann** = part, share, portion, division, area, point, district, region, and a similar word, **raon**, is a plain, field, meadow or holding. Thus we have Rhynie in Aberdeenshire, a wee share, Ranfurly, Renfrewshire, a farthing share, from *feòirlinn*, a farthing (a fourth of an old penny) or maybe the Rinns of Islay and Galloway, while Roinn Eorpa means Europe. See also **earrann** and **rinn**.

Taigh / tigh (pron. toy / tie, *tig* in Brittonic, *ty* in Welsh) often appears as **tay, ty** = house or hall, as in Taynuilt, house on the burn, from *uillt*, of the stream, genitive of *allt*, Tayvallich, house of the pass, from *bealach*, a pass between hills, as in Balloch, Tigh na Bruaich, house on the bank, from *bruaich*, of the bank, Tyndrum, house on the ridge, from *druim*, ridge or back.

Tobar = well or spring, as in Tobermory, Mull, well of Mary (*Moire* in Gaelic), Tibbermore, Perth, big well or Mary's well, while in Auchentiber, Ayrshire, and Dalintober Street, Glasgow, you would have once found a well in a meadow. **Fuaran** is another word for a well, or green spot, as in Tigh na Fuarain, house of the well.

Cultivation and Crop Names

(Grave accents have been inserted to indicate long vowels)

Field and meadow names feature a lot in place-names, especially *achadh*, *dail*, *blàr*, or *lèana*, plus the enclosure words *gàrradh*, *geàrraidh*, *goirtean*, *gort*, *gart* and *lios*, as listed above or *àirigh*, a sheiling, where there would be plenty *fodar*, fodder. There are also many words relating to animals and herds, especially *bò*, *bà* and *crodh*, cows and cattle, as in Bealach na Bà / nam Bò, Pass of the Cow / Cattle, or possibly Crow Road (Glasgow and Campsie Hills), while *sprèidh* also means cattle / livestock, and *tarbh*, *tairbh* means bull(s), as in Tarff, but *bàthach* / *bàthaich* is a byre, *buar*, a herd of cattle (also *treud* and *muthach*), and *muc*, *mucan* means pig(s), as in Auchtermuchty or the island of Muck.

We also find many crop names like *arbhar*, corn, *coirce*, oats, as in Achachork in Skye, or Corkerhill in Glasgow; *eòrna*, barley, as in Tìr an Eòrna, Tiree (see **tir**); *seagal*, rye, as in Glasgow's Auchenshuggle, a ryefield, and Balshagray, from *seagalach* (rye producing, plus *àirigh*, a sheiling or pasture), or *feur*, *feòir*, grass, hay or just *pòr(an)*, crop(s) in general. These would be often stored in a barn, *sabhal* (pron. sawal), as in Sabhal Beag and Sabhal Mòr, or *iodhlann*, a stackyard or cornyard, or *sgiobal*, a granary (maybe as in Skibo) and every family needed a handy *bràdh*, a quern, handmill or grinding stone, often appearing as *bràthainn* / *bràthan* (of the quern), as in the place the famous Brathan Seer came from, or for milling larger quantities, the necessary *muileann*, a mill of any kind.

Words relating to land use appear in *fearann*, farm, estate, fertile land, *buachar*, *innear* or *todhar*, all meaning dung, *ceapach*, tillage plot or land, in *treabhadh*, ploughing, and *crann*, a tree, mast or plough (or *crann-treabhaidh*), or in *clas*, a furrow, ditch (also *dig*, *dige*). *Goirtean* refers to a small patch of enclosed arable land, a park, *geat*, *gid*, is a smaller plot or lazy bed and *gart*, *gairt* is a field or cornfield, while a *glasach* is just a green field and *geàrraidh* refers to a garth, common grazing or pasture land, similar to *coitchinn*, *comp(h)airt*, common grazing, share, as in Kitchen and Cathkin and Cupar / Coupar, but some words just describe the size or number of the share, as in Coigach, fifth share.

The word *fasadh* or *fas* (see above) is a cattle stance or resting place, while various pen or fold words appear in *buaile*, a sheep or cattle fold, or *crò*, a pen, fold or circle, as in Linicro (fold meadow) on Skye, or *fàl*, a fold, circle, wall, hedge, turf or sod, as in Fall. A *fang* is a sheep pen (though *feannag* is a lazybed, or ridge of raised ground), while *buachaille* is a herdsman or shepherd, as in Buachaille Etive, the herdsman of Etive who would be looking after his *caoraich*, sheep, *uain*, lambs, or *gobhar* / *goibhrean*, goat(s), as in Bangour, West Lothian, from *beinn ghobhar*, peak of the goats.

There are also many references to butter (*ìm*), cheese (*càise*) and milk (*bainne* or *bainneach*, milky) in place-names, or to a milking pail or churn (*cuinneag*) as in A' Chuinneag, Assynt, but Butterstone and Buttergask in Perthshire and Butter Bridge in Argyll are bothersome places as they are possibly from *bothar*, an old word for a road, causeway or mud, though Ben Ime, butter mountain, is certainly near the bridge of the same name. Every settlement of course needed a *gobha*, or *goibhnean*, blacksmith(s) (pron. go-u, goinean) or a *ceàrd*, craftsman or smith (see **baile**), and a *ceàrdach*, a smithy, or a *teallach*, a forge, as in Challoch, Dumfries and Galloway, or an *innean*, an anvil, as in Inneans, Perth & Kinross, both places that had an anvil or forge, while a *sòrn*, a kiln or flue, could also be handy, as in the Ayrshire village of Sorn.

Farm or settlement names often feature words relating to position, such as auchter, from *uachdar*, top or high place, as in Auchtermuchty, upland place of the pigs or Auchterarder, upland of the high water (*àrd*, high, and *dobhar*, water) or even the Cumbric *ocel* / *uchel*, high, as in Ochil Hills, Ochiltree, Ogilvie and maybe Ogle or the opposite position with ìochdar, the lower, bottom part or place as in Iochdar in South Uist or Yoker, Glasgow, though Duntocher is from old Gaelic *tochar*, road causeway.

Flora

These are by no means exhaustive lists, only some frequently used examples. Trees, shrubs and plants (flora) often feature in Gaelic landscape, such as:

> abhall / abhaill (gen.) = an orchard or apple tree, as in Gartnavel, field of the apples, but can also mean a chief, as in Caisteil Abheil, the chief's castle, an Arran peak. See **ubhal** below
> aiteann, aitinn (gen.) = juniper, as in Tomatin, juniper hill
> bealaidh = broom, as in Ballater, broom land, with *tir*, land
> beithe = birch, as in Beith
> broighleag(an) = blueberry(ies)
> calltainn or calltuinn = hazel, as in Calton (pron. *caultin* and old Gaelic is *coll*, both close to the local pronunciation) and it also features in various Cowden(s) places
> canach, canaich (gen.) = cotton grass, bog cotton, as in Cannich river and glen
> caorann, caorainn = rowan, of rowans, as in Sròn a' Chaorainn, point of the rowans
> còinneach = moss
> conasg, conaisg (gen.) = whin, gorse
> craobh and coille = tree, wood, see above
> critheann, crithinn = aspen, of aspen, as in Killiecrankie or maybe Crianlarich
> cuilc(e), cuilcean = reed(s), bullrushes

cuileann = holly, as in Cullen
darach = oak, as in Craigendarroch
deanntagan / feanntagan = nettles
dearc = a berry, and dearcag = a little berry or currant
doire = a grove, usually oaks, as in places called Deer
dris, drisean (pl) = bramble(s), and driseag = a little
 thorn or bramble, as in Ardrishaig
droigheann, droighinn = blackthorn(s)
feàrn(a) = alder, as in Fearnan, Loch Tay or Ferniehirst,
 Aberdeenshire, or Fereneze, near Barrhead,
 Renfrewshire
feur, feòir (gen.) = grass, hay
fiodhag, fiodhaig (gen.) = gean tree, wild cherry
fraoch, fraoich (gen.) = heather, of heather, as in Freuchie
freumh, freumhan = root(s)
giuthas, giuthais = fir, pine tree(s), as in Kingussie or
 Carnoustie
iubhar, iubharan = yew(s) as in Inverurie and
 Tomnahurich, from *na h-iubhraich*, of the yew
leamhan = elm, as in Beinn Leamhain, Lennox and maybe
 Leven
learag (an) = larch
luachair = rushes, as in Leuchars or Locharbriggs in
 Dumfries and Galloway
lus, lusan = plant(s), herb, weed, wild flower, see **lios**
 above, often referring to herb gardens
muran, murain = marram grass, as in Eilean a' Mhurain,
 North Uist
preas, preasan = bush(es), shrub(s), thicket
raineach, rainich = fern(s), bracken as in Rannoch Moor
seamrag(an) = clover(s), wood sorrel, as in Beinn na
 Seamraig on Skye
seasg(a), seisge = sedge, reeds, rashes / rushes, barren
 and seasgann, marshy, as in Cessnock and Shiskine

seileach(an), seilich = willow(s) / of the willows, as in Auchnashellach, field of the willows
seileastair(ean) = yellow flag iris(es)
seòbhrag(an) / seòrach = primrose(s), maybe as in Glen Shurig on Arran
sgitheach, sgithich = hawthorn, of hawthorns
smeuran = brambles
sùbh(an) = (rasp)berry(ies)
ubhal, ùbhlan = apple(s) or as in *ubhal-ghort*, an orchard. See **abhall** above
uinnseann = ash

Fauna

Animal fish and bird names, both the domestic and the wild, also feature a lot, e.g.

bò, bà and crodh = cows and cattle, see above
bradan / ain = salmon
breac, bric = trout
broc, bruic = badger(s), as in Ibrox
calman / ain = dove(s)
capall, capaill = horse(s), a mare, *caipleach*, a place of horses
cat, cait = cat, as in Catacol on Arran, referring to wildcats, or in Lynchat, from *lann*
cearc, circe, cearcan = hen, of hen(s), chickens, cearc-fhraoich = red grouse
clamhan / ain = buzzard(s)
coileach, coilich = cockerel(s)
coinean / ein, coineanan = rabbit(s)
corra, corran = grey heron(s)
cù, coin = dog(s), cuilean(an) = pup(s), as in Conon and Contin (also **madadh** – see below)
cuthag / cubhag, cuthagan = cuckoo(s)

damh, daimh = ox, oxen or stag(s), as in Dava in Moray (and *àth*, a ford)

dòbhran / ain = otter(s), as in Craigendoran. Also biast-dhubh (black beast), madadh-donn (brown dog), madadh-uisge (water dog), muir-chù (sea dog)

druid, druide, druidean = starling, of starling(s)

each, eich = a horse, as in Eilean nan Each = horse island, or Lock Eck

eala(chan) = swan(s), as in Loch Nell, near Oban, from *nan eala*, of the swans

earb, earba = roe deer, of a roe deer

eun, eòin = bird(s), of birds, as in Culzean

faoileag(an) = seagull(s)

feadag(an) = plover(s)

feannag or starrag = crow

fiadh, fèidh = deer, of deer

fitheach, fithich = raven(s)

gèadh, geòidh = goose, geese

gòbhlan-gaoithe = swallow (forked bird of the wind)

guilbneach / nich = curlew(s), a *whaup* in Scots

iolaire = eagle, iolair-uisge = osprey, and there are 276 place-names in Scotland containing *iolaire* in one form or another, e.g. Benyellary in Galloway

lach(an) = wild duck(s)

làir, làire, làiridhean = mare, of a mare, mares

laogh, laoigh, laoghan = calf, of a calf, calves, as in Ben Lui

losgann, losgainn, losgannan = frog, of a frog, frogs

madadh, madaidh, madaidhean = dog, of a dog, dogs; wolf, of a wolf, wolves, e.g. Maud in Buchan

madadh-ruadh = a fox (also sionnach), and madadh-allaidh = wolf

maigheach, maighiche, maigheachan = hare, of a hare, hares

nathair, nathrach, nathraichan = snake, of a snake, snakes

partan, partain, partanan = crab, of a crab, crabs, adopted into Scots, as in Partanhall and others

piseag(an) = kitten(s)

ròn, ròin, ròin = seal, of a seal, seals, as in Eilean nan Ròn, Seal Island

seabhag, seabhaig, seabhagan = hawk, of a hawk, hawk, and speireag = sparrow hawk

searrach, searraich, searraich, foal (or colt / filly) of a foal, foals, as in Portinnisherrich, Loch Awe

sgadan, sgadain = herring, of a herring, herrings, as in Garscadden

sgarbh, sgairbh, sgairbh = cormorant, of a cormorant, cormorants. See also ON **skafr** or Scots *scart* or *scarf*

sionnach = fox, see **madadh-ruadh** above

sùlaire, sùlaire, sùlairean = gannet, of a gannet, gannets. See ON **sulan**

tarbh, tairbh, tairbh = bull, of a bull, bulls. See above

torc, tuirc, tuirc = wild boar, of a boar, boars, perhaps as in Brig o' Turk

tunnag, tunnaig, tunnagan = duck, of a duck, ducks

uiseag, uiseig, uiseagan = lark, of a lark, larks

An interesting class activity, especially in Gaelic-speaking areas, could be to collect examples of the above, especially those where no examples are given, or add other creatures to the list.

Strangely enough, in the Western Isles, the area we now consider the heartland of Gaelic, we find relatively few purely Gaelic place-names. The majority, especially in Lewis and Harris, are in fact of Norse or Gaelic-Norse origin, e.g. names beginning with H pop up all over the place, as in Hiort (St Kilda), or the mountains Heaval on Barra, or Hecla on

South Uist, while out of about 125 names for villages and crofts on Lewis, at least one hundred of them come from Norse. This Norse legacy probably explains their unique variety of Gaelic as the Norse incomers eventually intermarried and assimilated Gaelic culture, leaving a legacy of Gaelic words overlaying Norse ones and many Norse-Gaelic surnames, like MacLeod, MacAskill, MacDougall or MacIver.

While the Isle of Skye is full of Norse names, the island itself might derive its meaning from the Gaelic word *sgiathach*, winged (from its shape), but some have suggested it could also be from the Norse *sky*, meaning cloud, though Ptolemy in AD 150 refers to it as Ski or Skitis, long before Norsemen arrived. Even the Cuillin Mountains are not named after the legendary Celtic hero Cuchulainn and are probably not from the Gaelic *cuileann*, meaning holly (because of its jagged ridges), as in Cullen in Moray, but are more likely to derive their name from the Norse word *kiolan*, meaning high rocks or ridge. Thus many Norse names have been filtered through Gaelic before being inherited by English speakers.

9. NORSE – THE LANGUAGE OF THE VIKINGS

Although the Northern Isles (Orkney and Shetland) possibly had Norse settlers as early as the seventh century AD, they started arriving in large numbers from around AD 800 when Scandinavian countries embarked on huge waves of conquest and later just migration. Both Scotland and England were extensively settled by traders and farmers who didn't always come to plunder and capture slaves, even if they called places Brawl or Yell! Broadly speaking, Danes colonised north east England and Norsemen / Norwegians colonised the Northern and Western Isles as well as the northern mainland, but the east coast of Scotland did not provide safe anchorage and the natives were hostile, so they didn't hang around there. Although the word Norse is often used to refer to all these migrants, the term Scandinavian can also be used as an umbrella term to refer to all of them, whatever their origin.

As the Norsemen once controlled all the islands of the west coast as far as the Isle of Man, and the east coast of Ireland, Norse names appear right down our west coast into the Firth of Clyde as well as into Ayrshire and Galloway, an area settled by the Gall-Gaidheil, meaning the foreign Gaels, i.e. the Norse Gaels who were of mixed race, some from Ireland or the Isle of Man and who established settlements in the south west in the early tenth century. These Norse Gaels left a legacy of both Gaelic and Norse names, e.g. the Solway Firth (see **fjord** below), the River Ae (Norse *aa*, water) in Dumfries and Galloway or possibly Galston in Ayrshire, village of the strangers, though it could refer to Anglian incomers. Scandinavian names were of course also introduced to the south east and Border area by settlers from Northumbria who had an Anglo-Danish linguistic heritage.

There must therefore have been a lot of linguistic plurality or complexity in the south of Scotland around this time, involving Gaelic, Scandinavian, Anglic and later Norman English. In other words, in some places, people probably needed to speak, or at least understand, more than one language, or even sometimes to mix and match them by incorporating elements from different languages, something reflected in many hybrid or compound names in the south-east and south-west.

Norse influence in Scotland began to decline after the Battle of Largs in 1263 where the Scots under Alexander III sent them homewards to think again, at least according to the Scots. Nevertheless, the Norsemen still ruled over parts of the Western Isles for a long time and Orkney and Shetland for more than another two centuries and, as a result, their names predominate on these islands, especially the northern ones. In fact the Gaelic name for the Hebrides is Innse Gall, meaning 'islands of the strangers', which refers to the Norse settlers, and the Gaelic name for the Norsemen is Lochlannach / Lochlannaich (plural), meaning people of the loch lands or fjords. They often called themselves Ostmen / Austmen, i.e. east men, and they called the Gaels Vestmen, not men wearing vests, but west men.

As the Vikings were superb sailors with great skill in ship construction and navigation, many of our words for sea, island or coastal features, boats or nautical terms come from Old Norse, e.g. the Gaelic *birlinn*, a galley, or various ship places like Scalpay and Scapa Flow, from ON *scalp*, ship. For students who live in the Northern or Western Isles, it should be fairly easy to identify Norse names, not only for coastline, harbour or ship-related words, but settlement and farming words as well. However, Norse names can also be found in the south and west, especially in Galloway or even North Ayrshire.

Landscape and Coastal Features

Aa, aar, air = water or river (but *ar* is also found in Celtic river names) pop up in the consonantless Ae (Dumfries and Galloway), arguably Scotland's shortest mainland place-name, though maybe that distinction should go to the River E that runs into Loch Ness. This wee word also appears in the River Ardle and Strathardle (Perth), as well as maybe Edzell in Angus (with *dalr*) and Aros (Argyll), a river mouth (from *ar* and *oss*), as well as Thurso, possibly bull (*thjor*) river or Thor's river. A similar sounding word **ayre** / **ire**, a gravelly beach, bank, shingly spit, bar or tongue (from *eyrr* or *eyri*) appears in Northern Isles Ayre places, such as Point of Ayre, a meaning similar to *dòirlinn* in Gaelic and a *tombolo* in English (from Italian *tombolo*, a cushion or pillow).

Bekkr / **back** / **bakkr** = stream, river or valley through which it runs, and applied to a settlement on its banks, found in the Borders and south west, as in Allersbeck, Bodsbeck, Craigbeck, Greenbeck, Waterbeck, usually named after a physical feature of the landscape or people who lived there, often appearing in compound names where the describing element is Anglic / Scots and it is common in the north of England. Not to be confused with Gaelic *beag*, meaning little. See **gil** / **ghyll** below.

Brae, from ON *bra* and Gaelic *bràigh* = steep slope, bank, face of a hill, uplands, as in Braeheid / head, Braeside, Links Brae, Braes o Killiecrankie and braes of many other places or people. Most of the Lowland Scots braes probably come from the Anglo-Danish Northumbrians. **Bakki** is another Norse word for a bank, as in Coldbackie, on our north-west coast, but 'brae' in Shetland can have different meaning as it might come from the ON word *breiðr* meaning broad.

Dalr = dale or valley but later came to imply a settlement in a valley and appears in hundreds of places. If used as a suffix (after the name) usually it's Norse, e.g. Armadale on Skye which is probably from the Norse *arm-r dalr*, an arm-shaped valley, Berriedale, Beri's dale, Borrodale, fort dale, Helmsdale, meaning Hjalmund's dale, various Swardales (from *svard*, a grassy slope or sward) or Rodel / Rodil, roe deer valley, or Saddell in Kintyre, possibly from *sag dalr* (saw dale). Many Gaelic dale names, especially on the Western Isles, come from *dalr*, though southern dales are from the English cognate, as in Clydesdale or Liddesdale. As you can see, it is often named after the people who lived there or named after a river and is very common in the north of England, as in Yorkshire Dales. The Cumbric and Gaelic equivalents are *dol* and *dail* which are used as prefixes.

Ey = island and usually appears in the last element of a word as **ae**, **ay**, though **eidh** is an isthmus, as in the Eye Peninsula, Lewis, Loch Eye in Sutherland, or several **aith** places in Shetland. The extent of the Norse influence is shown by the fact that the vast majority of Scotland's nine hundred or so islands end in this word. One of the Gaelic names for Iona, 'I' or 'Idhe', comes from this word, the former being one of the shortest names in the world! Other examples, often named after people, fauna or position are: Bernera, meaning Bjorn's island, Eriskay, Eric's island, Islay, Ile's island (and people from there are still called Ileachs), Jura, possibly meaning Doirad's island, or alternatively from *dyr* and *ey*, deer island; Oransay is perhaps Oran's island, Pabbay, priest's island, Rousay and Ronaldsay, Rolf and Ronald's islands, while Rothesay is Roderick's isle. Pladda, Fladda, Fladdey and Flodda are just flat islands, Foula is bird island, Orkney (called Orc in earlier times) is boar or pig island, possibly after the Boar tribe, Raasay, roe deer island (from *rar*) and Whalsay, the whale island. Scalpay is a boat-shaped island, from *skalp*, skiff, ship, but Flotta is the fleet island (from *flotr*), Mingulay is a big island (but *mikla*, big, has been transposed into mingil), while Hoy is a high or tall island, from *ha*, tall. Skye is possibly the winged or divided island, from Gaelic *sgiathach*, winged (Skye folk are Sgitheanachs), Vatersay is the water (i.e. fresh water) or wet (from *vatn* or *vatr*) island, a damp place, and Westray is just what it says: west island. There are of course hundreds more, some less obvious, such as Harris which means high island, from *har*, high place, and *ey* (rather like Hoy, above), or Texa, off the coast of Islay, which has nothing to do with the American state of Texas, but is probably from the Norse *t-heggs*, bird cherry, but maybe also from Old Gaelic / Irish *tech*, an earlier form of *taigh*, a house.

Fell or **val**, from *fjall* / *fjalr* = mountain, hill, fell, as in Campsie Fells (see G **camas**), Bleaval on Skye, possibly meaning slatey blue mountain, Criffel, near Kircudbright, a split fell, from *kryfya*, split, Goat Fell on Arran, goat mountain, Heaval on Barra, a high mountain, but disguised in Arkle in Sutherland (Arcuil in Gaelic), an arc-like mountain, from *arkfjall*. The word has sometimes gone through a linguistic transformation process and is well disguised, as in Ben Loyal, from *laga fjall*, meaning law hill (i.e. keeping the law) or maybe from *leidh*, a levy or muster of soldiers, so it could be a muster mountain but not a mustard mountain! Scaefell and Snaefell in England are a scrape and a snow mountain.

Fjordyr (fjord) = firth, river estuary, sea loch, or even a bay, as in many of our sea lochs, though some come from the Anglo-Danish cognate, **firth**, e.g. Firth of Clyde, Firth of Forth or the Solway Firth, the latter being the firth of the muddy ford, from Norse *sol* (mud) and *vath* (ford). It is sometimes concealed in places like Minard in Argyll, a small bay from ON *minni*, small. At the opposite end of the scale is a **gja** / **geo** / **goe** (*geodha* in Gaelic), a sea cleft, ravine, creek, a steep narrow inlet, as in Papigoe, near Wick, from

Norse *papa*, priest's creek. **Creek** is from the Norse *kriki*, meaning a corner or neuk and OE *creke*, a narrow inlet on the coast, a place you don't want to go up without a paddle! **Krokr** is another similar sounding word, meaning a bend, as in Crook of Devon, Perthshire, and some crooked places in Ayrshire, but derived from Anglic rather than Norse.

Gil(l) / ghyll = a ravine, gully, narrow valley or stream and found a lot in the Border area, as well as the Northern Isles, but we also find Corrygills on Arran, probably from ON *karri*, cock ptarmigan, and *gill*, so the gully of the ptarmigan. Also see **bekkr**.

Holm / holmr = a small grassy island / islet, often used for pasture, as in many holm places in Orkney and Shetland, such as Glimps Holm. See also OE / Anglic section.

Hop / hope = a bay, usually a shallow one, as in Longhope, a long bay, or shortened to **òb, oba** or **tòb** in Gaelic, from which we get Obbe on Harris, Oban, a little bay and Opinan, a little sheltered bay in Ross-shire. **Ham** and **hamn (ON hofn)** are other harbour or haven words, as in Hamnavoe, harbour of the bay, and could refer to a settlement. See **vik** below.

Kames, from ON *kambre* = a crest or mound, as in Kames of Hoy and many other northern kames places, especially on the Orkney and Shetland Isles. See also Old English section.

Klettr = rocky holm or cliffs, as in Clett in Orkney and Shetland, Clatt in Aberdeenshire, and maybe Clatterin Brig in Angus and Clatteringshaws in Galloway. See also **hus** and **clett / cleat**. **Klakk** = a rock, as in Clickhimin in Shetland and elsewhere, but a **knappr** is a nob or lump, as in various Knapp places or Knapdale, Argyll, where you always have to watch your napper! Coll is possibly derived from **kollr**, a bare head or top, so a bald place, while **hella / hellya** means a flat rock, but not as in Up Helly Aa, the famous Shetland mid-winter fire festival.

Law on the other hand has nothing to do with lawmen but is simply a conical hill or mound, as in Sidlaw, possibly meaning hill pasture, from Norse *saetr*, though Lowland laws, like Berwick Law, are mostly from the Old English *hlaw*.

Megin = great, **mikla** or **mikil** (as in Scots *mickle / muckle*) = big and **minni** = small, appearing in Norse names such as Minard, little bay or fjord, the Minch, maybe a great headland, from *megin* and *ness*, Mingary, a big enclosure, from *mikla* transposed to *mengil* and *gardr*, enclosure, or Mingulay, a big island from *mikla / mingil* and *ey*, island. **Storr** can also mean great, as in Papa Stour in Shetland, the great priest island, or the Storr on Skye, while **peerie** and **peedie** can refer to anything small, including places, in Orkney and Shetland.

Mire = bog, swamp, from ON *myrr*, a mire, as in OE *mos* (passing into Scots as **moss**) e.g. Mid Mire and Rossmire in Orkney. Tam O' Shanter 'skelpit on thro' dub and mire.'

Nes = headland, usually as the last element in a word, as **ness, nis, nish**, e.g. Aignish, ridge point, Callanish, Kali's headland, a personal name, Bo'ness, a shortened form of the name Borrowstounness, a headland farm belonging to a man called Beornweard, or Durness, deer cape, point, from *dyr*, deer. Stromness means cape in the current, from *straumr*, Stenness, rocky cape, from *steinn*, stone, Trotternish, headland of Throndar, a personal name, Waternish, from ON *vatn*, water cape (as in Vatersay), while The Minch may refer to a great headland (Cape Wrath or Butt of Lewis) from *megin*, great and *nes*. Caithness is maybe the cape of the cats or cat men or tribe, from Gaelic *cataibh* (pron. caitiv), but it may have been a Pictish province or even the personal name Cat or Cait, the son of Cruithne, supposedly the 'founder' of the Picts. We find *ness* in a great many other places in the Northern and Western Isles, as well as in Dungeness and Skegness in England.

Skari = shore, as in Skara Brae, an example of a Scots word added to a Norse one, but **skarpr**, shortened to **scarp**, means barren, or steep cliffs, as in the island of Scarp, though it might also refer to the *skarfr*, or *sgarf* / *sgarth*, the cormorant or shag and you will often find them on the many skerries round our coast. We also find **sker, skar, skaw**, from *skjaer* (or Gaelic *sgeir*, giving us skerry / skerries in Scots / English), a sharp rock, rocky isle or reef, as in Talisker, sloping rock from *t-hallr*, sloping, or Scrabster, a rocky farmstead from *ster*, as in farm (see **bolstadr** below). The coastline of the north and west is covered in them. **Skeri** can also mean cutting, as in cutting peat. **Skal**, though, refers to soft rock, appearing as *skel* in Skelbister, a farm of soft rock. Similar sounding are **skagr** or **skogr**, wood, as in Scourie, and **sag**, a saw, as in Saddell.

Stav, stafr = staff, stick, stave, post, pillar, as in the island of Staffa or Staffin on Skye.

Thang, tunga / tangi = a low headland, a spit of land or tongue, as in Taing and various places in the Northern Isles (see **ting** below and Gaelic **teanga**), but the Old Norse word **hofud**, a headland, is usually just a head and **horn** is another cape word, as in Kishorn in Wester Ross which really sticks out, as it is from ON *keisa*, protruding.

Vik = creek or bay, usually appears as **wick, vig, vaig, aig, uig**, e.g. as in Wick, Uig, or Nigg, thought to be derived from *an ùig*, a Gaelicised version of *vik*. Some are very well disguised indeed, like Oldshoremore in Sutherland which, believe it or not, comes from Asleifar vik, Asleif's bay, with the Gaelic *mòr*, big, added later to emphasise its size. **Vagr** (usually appearing as **way** or **wall** or **voe**) is also a bay or inlet, as in Kirkwall, Walls, Pierowall (maybe a peerie / peedie or wee bay), Sullom Voe, the gannet bay, and Hamnavoe (see **hamn** above) in Orkney and Shetland. We have hundreds of them from Arran to the Shetland Islands, as in Arisaig and Aros bay, from Gaelic *àros*, a house or habitation, now a place-name, but also a river mouth, Brodick, broad bay, from Old Norse *breithr*, broad, Diabeg, a deep bay, from *djup*, deep, Kirkwall or Kirkhope, both meaning church bay, Lerwick, mud bay, from *leir*, mud or clay, so no wonder other Shetlanders often thought 'Lerrickans' were a mucky lot. Mallaig is a headland or shingle bay, from *muli* or *mol*, Sandwick and Sannox are sandy bays, the latter from Gaelic Sannaig via Norse Sandvik. Scalloway is a sheiling or shelter bay, from *skali*, shelter or hut, Scavaig is claw bay, possibly from *ska*, to scrape, and Stornoway is a steerage bay, from *stjorn*. Reykjavik, the capital of Iceland, means misty or smoky bay, as in the Scots word *reek*. See also **hope** above. However, the OE / Anglic *wic* or *wick* means a place or dwelling and not a bay, as in Hawick, while the Anglic *hop* is a hollow which gives us the Scots *howe* (see below).

Settlement Names

Bolstadr / bolstathr = farmstead, estate, from *bol*, a place or share (same as Pit) and *stadr*, a farm (OE / Anglic *stede* / *stead*). **Stadir** or **stathir,** sometimes shortened to **sta,** appears in the very earliest settlements and often had high status, e.g. Gunnista, Gunni's farm, Olistadh, Oli's farm. Bolstadr can be found in various forms or disguises in Orkney, Shetland, the Western Isles and the northern mainland, often named after people, animals or location, and appearing in dozens of places with different endings:

> **bo**, e.g. Skibo = ship place or **bost(a), bist**, e.g. Carbost = farmstead by the copse (little wood), Habost, high farm, Kirkibost, church farm, Leurbost, clay farm, Shawbost, farm by the sea loch, from Norse *sja*, or Boust on the island of Coll.

pool, pol or **boll, bull**, e.g. Ullapool, Olaf's farmstead, Eriboll, farm at the gravel beach or Bayble, not the biblical tower of Babel, but the priest's farm, from *papa* and *bol*, while it also appears as *bull* in the form of Bull of Hove or Bull of Kerston in Orkney.

setr or **saetr, shader** = sheiling, cow pasture or farm, both originally meaning temporary dwellings or shelters, but *setr* came to mean a dwelling that was a bit more permanent and it is very common in the Northern Isles, as in Melsetter, Mossetter, Vatsetter, or Setter, Shader, or Uigshader, bay sheiling or pasture and Marishader, mare pasture, on Skye.

ster, e.g. Isbister = easterly farmstead, Kirkbister, church farmstead, Scrabster, rocky farm, from *sker*, a rocky isle or reef or skerry, or Lybster, a lee or sheltered farm, from ON *hlie*.

Bru / bru'r = bridge, as in Brogar, a bridge enclosure, from *bru* and *gardr*, or Brora and maybe Bruar, a bridge river, from *aa*, river. **Bryggja** also = bridge, giving us OE / Scots *brig*.

Bur = a house or shed, and **bu** (also OE) = a homestead, or livestock, cattle, as in Bower (Highland). The Gaelic word *buar* means cattle, but the Scots **bour** or **bower** refers to a tenant who hires cattle and has grazing rights, a bouman, a man in charge of cattle on a farm, a tenant with a *bow* or bowhoose / house, a cattle shed, as in Bowhouse (Scotland and England). See Scots farm words.

Burra, borg = fortification as in Borve or Borgue, from *borgar*, fortified places in the Western Isles, Skye and Galloway, or Brogaig, from *borgvik*, fort or castle bay, or Borrodale, the dale of the fort, or West Burra and Noust of Burraland, Shetland. See OE / Anglic **burgh** or **burh**.

Byr = farm, village or hamlet, like *bolstadr*, and usually appears as **by** or **bie** and often in compound names with

Anglic or even Gaelic elements. It is common in the Borders and south west and, like many places in the north of England, from Anglo-Danish, such as Whitby, white farm or village. In Scotland we have quite a few places named Busby or bush farm or Crosby, cross farm or village, as well as Canonbie, the canons' hamlet or village, and farms named after their owner, such as Duncansby, Donald's farm, from Norse Dungal, Dounby, fort farm, Lockerbie, Lockhart's village. Sorbie (or Sowerby in Yorkshire) probably indicates a muddy, boggy (from *saur*), hence a sour or difficult farm, though other meanings are possible, while Netherby is the lower farm, from Norse *nedri*.

Clett / cleat (see **klettr**) refers to a stone built house, possibly a church, e.g. Nether Benziecleat (maybe from a personal name, such as Benti, a diminutive of Benteinn, or from *bein*, bone, or *beiner*, a help or benefit) in Orkney, or the *cleits*, stone bothies or storage huts, found on St Kilda and its neighbouring islands. It also features in cleat or cleiteadh, (the Gaelic version), rocky places around the north and west coast, such as in the south coast of Arran, so it can refer to coastal features as well as settlement.

Gardr / garðr, a Norse word similar to Gaelic words for enclosure, garth, field, as in some **girth** or **garth** places, especially in the south west, or Biggar, barley field, from Norse *bygg*, barley, or Rogart, red field, from *raudr*, or Calgary, Kali's (personal name) garth, or Mingarry, where *mikla*, big, has evolved into *mingil gardr*, an enclosure between machair and moor in the Western Isles. The Old English word is *geard*, as in Scots *gairden*. See also Gaelic **geàrraidh** and **gart**.

Hus / housa = house, a common Germanic word, as in Old English, usually fairly obvious in many Scots names, but even when anglified to house, still often pronounced locally as hoose, as in Crosshoose, Ayrshire. It is sometimes well disguised as in Lewis which is possibly from the Norse *ljod / lydr* = people or common folk (but *lydr* can also mean music and song) and *hus*, thus meaning people's house and via the Gaelic Leodhas to Lewis, though it has also been linked or confused with Gaelic *leoghuis*, marshiness.

Kirkja = church, usually appears as **Kirk**, a common Norse and Germanic word, found throughout Scandinavia, Holland and Germany and became the Scots and northern English form which southern English changed to church. Thus in kirk names of Norse origin we have Kirkwall, church bay or Kirkness, church cape, and others indicating location, but Kirkoswald in Ayrshire, meaning Oswald's church (where Tam o Shanter's 'ancient trusty drouthy cronie', Souter Johnnie, came from) is from Old English. See English **circe** or Scots **kirk**.

Ky-r = cow(s), as in the Scots word **kye**, from which we get Cursetter in Orkney, from *ky-r* and *saetr*, farm or pasture land and maybe the surname Cursiter / Cusiter. A related word is **kvi**, a cattle pen or fold, church, found as **quoy** in Shetland and Orkney as in Quoyburray and Quoyloo, or Quinish, a cattle fold headland in the Western Isles and Quiraing, a crooked enclosure on Skye, from *rong*, crooked, but nothing to do with a lack of honesty. It appears in Gaelic as *cuithe* which features in Arran's Cuithe and Cuithe Meadhonach, a middle cattle-fold.

Midden / middin, from ON *myki dyngja* = a dunghill, midden or boggy place (also OE and Scots) or a boggy or smelly place, as in a number of Midden or Middens places, all of which are definitely right middens!

Saetr = pasture land, shieling, farm, maybe as in Sidlaw, while **svard** or **sword** has nothing to do with weapons but refers to a grassy slope or sward, as in Swardale or Swardland. See also **bolstadr** above.

Thwaite, twatt, that, at, from *thveit* or Danish *thwet* = clearing, meadow, paddock, a word found in field and settlement names, often referring to animals or terrain, such as various *twatt*s in Orkney and Shetland. In the south west we find Cowthat, Murraythwaite, Murthat and Thwaite in Dumfries and Galloway, or Macherquhat in the Stinchar Valley, combining the Gaelic *machair* and Norse *thwaite*, meaning a meadow field, another tautological name and evidence of the Norse Gaels' legacy in that corner of Scotland. Also found in England, especially in the north west and the Lake District, and in Normandy from the Danish or Norse incomers.

Ting / Thingvollr = meeting ground or field, assembly or parliament, as in Dingwall in the Highlands, Tinwald in Dumfries and Galloway and Tynwald, the name of the Manx Parliament, all related to *tunga / tangi*, a spit or tongue of land, as in Tongue on the north coast and Tongland in Galloway and possibly Changue Hill in the Stinchar Valley, though more likely to be from Gaelic *teanga*, a tongue. Assemblies are often places for people to exercise their tongues. The Old English word is *tonge*.

Vollr = field, meadow or open space, as in Thingvollr, and it features in some *wall* places, like Dingwall and Tinwald above, or maybe in Sheanawally Point on the Wee Cumbrae.

Norse Flora and Fauna

More words can be found on various websites, but here are some common examples:

 busk = bush, as in Busby, i.e. bush farm
 faer = sheep, as in Fair Isle and Faeroe Islands
 fugl, foul = fowl, bird, as in Fitful Head, Shetland, from *fugl* and *fit*, foot
 hross = horse, mare

ky-r = cow(s), see above, as in the Scots word *kye*
lax = salmon, as in Laxford, Sutherland
lin = flax, *lìon* in Gaelic
ling, lyng = heather
mar, ma = seagull, as in the Isle of May or Manish in Harris, gull ness
mosi = moss, as in Mousa, Shetland
orn or ari = an eagle, possibly as in Arnol on Lewis
ra, rar = roe deer, as in Rodel, Harris and the island of Raasay
rauor, rawdr = red, as in Rogart, Sutherland
reynis = rowan, as in Lochranza on Arran, loch of the rowan river, from *reynis* and *aa*, river
seil = seal, as in Seil island
sil, sild = herring, as in Shieldaig, herring bay
skarfr, or sgarf / sgarth = cormorant or shag, *sgarbh* in Gaelic, *scart* or *scarf* in Scots
sulan = gannet (*sùlaire* in Gaelic and *solan goose* in Scots) as in Sullom Voe or gannet bay
thorskr = cod, possibly as in Tarskavaig on Skye, meaning cod bay
whin, quhin, fun = gorse, whin (also OE and Scots)

10. OLD ENGLISH / ANGLIC / SCOTS

The Angles, Saxons and Jutes were Germanic-speaking tribes from northern Europe who began to settle in England from about the fifth century AD, especially after the Romans left. The Jutes and Saxons settled in the south east of England but the Angles settled mainly in the north east and for a century or so also controlled parts of southern Scotland from around the seventh century onwards, including the Lothians, Galloway and even parts of South Ayrshire, leaving a linguistic legacy on some place-names or relics like the Ruthwell Cross. By the mid-seventh century the Firth of Forth marked the northern boundary of their Kingdom of Northumbria, though further northern expansion was halted by the Picts at the Battle of Nechtansmere or Dunnichen in 685.

However, the later arrival in the north of England of Norsemen, especially Danes who used a closely related language, created a mixture of Anglic and Danish, often referred to as Northumbrian, a language that had a huge impact on our linguistic heritage. Indeed, many of the words that we think of today as typically Scots, such as auld, brig, burn, hoose / hous, are in fact from this Anglo-Danish Northumbrian tongue.

In the wake of the Norman Conquest of 1066, this Northumbrian dialect spread extensively into Lowland Scotland, gradually ousting northern Brittonic / Cumbric from the Borders as well as the south east and west. From around the twelfth and thirteenth centuries onwards, Gaelic was gradually pushed further north, until Anglic or 'Inglis' became the dominant language in the Lowlands as well as the north east, especially with the setting up of burghs to promote commerce, where 'Inglis' became the lingua franca (the common 'lingo' of trade). Due to our 'auld alliance' with France and our trade with the Low Countries, many French, Flemish and Dutch words were also absorbed into the mixter-maxter tongue that eventually become known as 'Scots'.

The term 'Scottis' originally meant Gaelic, but by the fifteenth century the English spoken in Scotland was being referred to as 'Scottis' to distinguish it from southern English because it was now being seen as a distinctively different dialect or language and it gradually became the language of the court, literature and law, spoken by all classes in the Lowlands, north east and Northern Isles. Unfortunately, the status of the Scots language began to decline after the Unions of the Crowns (1603) and Parliaments (1707) and the use of the English translations of the Bible after the Protestant Reformation, from the late sixteenth century onwards.

Another serious blow to Scots came during the eighteenth and nineteenth centuries when the Scottish upper classes increasingly looked on Scots as inferior or 'incorrect' English, something from which the language has not yet fully recovered, especially as this attitude took deep root in our educational system.

Although Scots still suffers from this legacy of ignorance and prejudice, there is generally a more positive attitude towards the language today. There is now a national qualification in Scots, and, at the time of writing, there is a proposal to give it equal legal status with English and Gaelic via a Scottish Languages Bill at Holyrood. It is in fact a language with a wide variety of dialect variations from Shetland to the Borders as well as across the Irish Sea, where we have Ulster Scots which came from Scots-speaking settlers (especially during the Plantation of Ulster in the seventeenth century), a language possessing its own rich poetry and song traditions.

Important place-names in the Scottish Lowlands, such as those of settlements, parishes and major landscape features, are mainly of Celtic origin but many Cumbric and Gaelic words were incorporated into English and Scots names. Scots also assimilated many Gaelic loan words or in some cases translated the Gaelic meaning into Scots, possibly during a period when both tongues co-existed, e.g. Ruchil(l), or rough hill in Scots, probably comes from the Gaelic *garbhach*, a rough place, as in Garrioch and Garry places.

Probably more Scots place-names appear on detailed maps because most of the surviving names of the minor landscape features, as well as later settlements, were given their present names by Anglic / Scots speakers from the thirteenth and fourteenth centuries onwards, often with 'affixes' attached to them to describe the various divisions, e.g. west(er), east(er), nether, laigh or low, heich or high, meikle, muckle, big, and so on, mostly attached to place-names that are much earlier. You will find more detail about this if you look up the Ordnance Survey's guide to the Scots origins of place-names.

It might be interesting for students, not only to trace some of the common Anglic / Scots words for landscape and settlement in their area, but to compare the number or percentage of these with Celtic or Norse ones and maybe even investigate the historical reasons for this. It might also be possible to investigate or trace on the map where older Scots words have been changed into their modern English equivalent, such as hoose / hous to house, auld to old, etc., and to discuss the reasons for this and maybe redraw the map to show the older names.

Landscape Features

Bent = coarse or reedy grass on hillsides or moorlands, i.e. marram grass, from OE *beonet*, as in Bentheads, Bentfoots, Bents and other bent places.

Brae, from ON *bra* and Gaelic *bràigh* = steep slope, bank, face of a hill, uplands, as in Braeheid / head, Braeside, Links Brae, Braes o Killiecrankie and braes of many other places or people. Most of the Lowland Scots braes probably come from the Anglo-Danish Northumbrians.

Burn, from OE *burna* = a stream, brook and is by far the largest group in Scottish-English stream names (at least 2,650 on the one-inch Ordinance Survey map), found all over Scotland in one form or another, mainly with the generic term *burn* featuring second, as in Bannockburn, little white stream, from Brittonic *ban* and *oc* (or from *bannauc*, a peaked hill), but over 260 names have the defining element coming after burn (usually referring to a landscape feature or settlement) and linked to it by the preposition *of*, as in Burn of Glendui or Burn of Drumcairn, a structure derived from the Gaelic Allt na ... (see Gaelic **allt**). Most towns or villages have at least one burn, often several, sometimes with the burn word added to a much older water name, as in the Putyan, Paduff, Pundeavon or Powgree Burns of the Garnock Valley in North Ayrshire. *Bùrn* is also the Norse and the Gaelic word for fresh water. A larger burn, but not usually as big as a river, is sometimes referred to as the water of, as in the Water of Leith, Water of Tulla, Water of App or Water of Ken, which again reveals an older Gaelic substructure in spite of having been anglified, while in other places the word burn has also been anglified to water, as in Gogo Water, or Dusk Water. The OE word *broc*, a brook, can also appear in burn names, as in Broughton, a settlement by the brook.

Cleugh, cleuch, from OE *cloghe* = a ravine, glen, gorge, cliff or crag with steep sides, as in Ben Cleuch, Byreclough, Buccleuch, from *buk*, a buck, or Cleugh Hill and many other cleuch names in southern Scotland.

Dale = a valley and, if used as a suffix, usually it's from Anglic as in Clydesdale, Tweeddale, Nithsdale, Alandale, or Norse as in Helmsdale, Borrodale, often named after rivers, as in Tweeddale, or named after the people who lived there. It's very common in the north of England, as in Yorkshire Dales. Gaelic *dail* is usually a prefix. See also Norse **dalr**.

Den / dean (OE *denu*) = a hollow, dell, valley, ravine, often used in the Lowlands instead of glen, as in Aikendean, oak den, Dean Village, Den Burn, Denholm, Dennyholm (see **holm** below), Lambden, Cardenden (see Brittonic).

Dod = a bald or rounded hill, from OE *dodden*, to clip or poll, as in Windy Dod, Dod Law (a hill hill) or various Dod Hills in Ayrshire, Dumfries and Galloway and the Borders, a word often used in Scots (a *daud*) to mean a lump or chunk of something, as in 'giez a dod o'.

Dubbs places are puddly places, from S *dub*, a puddle or pool, a word probably derived via OE from Low (northern) German *dobbe*. We find them in Stevenston, Beith (also Neilson), and in the Dubs Burn, south of Darvel in Ayrshire, though Br *dub* / Gaelic *dubh*, black, might also be present in some, suggesting a place of mucky black pools. Tam o Shanter knew plenty about dubs as he 'skelpit on thro' dub and mire'.

Edge / ege = crest, edge, sharp ridge, as in Edgefield or fauld, Cairnridge and many Windyedges or Windyridges, all good places for windmills.

Haugh, hauch = meadow or flat land beside a river, found in many places, e.g. Carterhaugh, Spittal Haugh (refuge or hospital meadow), Rosehaugh, Haughheid / head, Haughs o Cromdale, or Philiphaugh which didn't belong to Philip, but is from OE *ful*, closed and *hop*, a valley or hollow, though Ballyhaugh on Islay is probably from the Norse *haugr*, a mound.

Heuch, heugh = pit, cliff, steep bank, overhang, glen, ravine (like *cleuch*), as in Coalheugh, Ravensheugh, Redheugh(s), Slackheugh, Slateheugh. The poet Hugh MacDiarmid, born in Langholm, wrote, 'There's teuch sauchs growin i' the

Reuch Heuch Hauch', an apt description for many parts of the Borders.

Holm = meadow, flat land beside a river or islet, a word common to Anglic and Norse, found in many parts of the Lowlands, including North Ayrshire, e.g. Holmbyre, a meadow farm, near Dalry, Dennyholm, Kilbirnie, or Bartonholm, an old mining village near Irvine, meaning meadow of the Britons, or Groatholm, a meadow of hulled or husked grain, especially oats, or Kirkholm in the Stinchar Valley. See also Norse section.

Howe = a hole, hollow, low lying area, basin, depression, from OE *hol*, found in many Scottish low places, such as the Howes of the Mearns, Fife or Buchan or Howgate in West Lothian, though some *how* names in the Western or Northern Isles are from the Norse word *haugr*, a mound, heap or midden, as in Maeshowe in Orkney, or Howmore in South Uist. **Hop / hope** can also mean a hollow or valley in Anglic, as in Wauchhope, from *walc / wealh*, an outsider or foreigner, or Dryhope, from OE *dryg*, which could be either a dry hollow or a fort one. See also **kerse** and **merse**.

Kame / Kames (from OE *camb*) = a small mound, a ridge, or a small peninsula or isthmus, as in many places called Kaimes, Kaimhill (a hill hill), Kaimend, Kaimflat, Kaim Head, or Kames of various places. However, some kame(s) places are from the Gaelic *camas*, a bay, as in Kames Bay on the Big Cumbrae, or from Gaelic *cam*, bent or crooked, while those in the north and Northern Isles are from the related ON *kambre*, a crest or mound, as in Kames of Hoy. See also Norse section.

Kerse, carse = low lying land beside a river, as in Kerse, Kersie, Kersland, Carse of Gowrie, often used in farm names, though in former Norse areas this Scots word may be from ON *kjarr / kerss* with the same meaning, which passed into Gaelic as *càr(r) / càir*, meaning a marsh or mossy or fertile plain, and we are not short of such soggy places. See **merse** below.

Knowe = hill, hillock, very common in Lowland Scotland, as in Robert Burns' song 'Ca the Yowes tae the Knowes', from OE *cnol*, a knoll. Wee knowes pop up all over, e.g. Cowdenknowes, a place commemorated in another famous folk song, meaning hazel hillocks or knolls, from Gaelic *calltainn*, hazel.

Law = a conical hill or mound, from OE *hlaw*, but also Old Norse, e.g Berwick Law, Black Law, Dod Law, Irish Law, Harlaw (hare hill, from OE *hara*), Mintlaw, thus a land not short of laws, especially Wardlaws or Ward Hills, which were sentry or lookout places, often with a beacon, from OE *weard* and ON *vardr*, a watch. We also have the pleonastic, but not fantastic, tautological name of Law Hill, another hill hill!

Links, linkis, from OE *hlinc* = rough, sandy grassland, including dunes, along the coast, often common land belonging to a village or town, much used for gowf / golf and probably where the game started, though this is much disputed, e.g. Leith Links, Linksfield or Lundin Links in Fife which has nothing to do with the city of London, but is an old Pictish name related to Gaelic *lodan*, a marsh. The Ayrshire coast has almost no end of links courses.

Mersc, mersch, merse = marsh, marshland, flat alluvial land beside a river or estuary, as in the Merse, Berwickshire, which, along with kerse, explains why we often get stuck in the mud. **Misk** is another word for a damp, boggy low-lying place, as in Misk names, such as Misk Knowe in Stevenston, Ayrshire.

Moor, muir, from OE *mór*, as in Boroughmuir, moor of the burgh, or Morton, moor farm. Moor / muir has often replaced Gaelic *mòr* (big), as in Dalmuir which should be

Dalmòr, while some *auld* muirs might go back to G *allt*, burn muirs. Also **moss, mos** = marsh, bog, from which peats were cut for fuel, as in various moss places, and can be found as a prefix or a suffix, such as Mosspark, Mosshead / heid or Airds Moss.

Neuk, nuke = a nook, corner, also a promontory, inlet or outlying place (from *noke* of the Middle English period, i.e. between the twelfth and fifteenth centuries) as in the East Neuk of Fife or the ingle neuk, the fireside corner, where Tam o Shanter sat in the pub in Ayr.

Rig, ryg = ridge, from *hrycg*, a common Germanic word, as in the Norse word *hryggr*, but most Scots 'rigs' are probably from the Anglo-Danish incomers, e.g. Lanrig, which possibly means lang rig / long ridge (but also possibly from Brittonic *llanerch*, a clearing, as in Lanark), Longrigend, long ridge end or Lintseedrig (pronounced Linsyrig, a farm near Dalry, North Ayrshire) meaning ridge of the lint seed. A ryg or rig is also a common Scots farming term from the Run Rig system of agriculture, as the land was cultivated in raised strips or ridges, with drainage channels in between, the remains of which can still be seen on many hillsides. Rig can therefore often refer to settlements.

Scrog, scrogg = scrub, brushwood, thicket, stunted bush, stump or root (from Middle English *skrogg*), as in Scroggs (Dumfries) or Scrogbank (Selkirk).

Seggs, seggy = sedges, sedgy, covered with sedges, marshy, as in Seggiecrook, Banff, Seggy Neuk, Kirkcudbright, Seggiehill, Fife and the old Kilwinning name, Segdoune / Segton. It also gives us the surname Seggie.

Shaw, schaw (from OE *sceaga* and ON *skagi*) = small wood, copse, grove, as in Hareshaw, meaning either hare wood or high wood, from Scots *har* / OE *hiera*, high, or Pollokshaws (see **pol**), Shawfield, Wishaw, a willow wood, from OE *withig*, or Shawwood, near Catrine in Ayrshire, a wood wood, another place-name tautology, or the House of Shaws in Robert Louis Stevenson's novel *Kidnapped*.

Sheuch, sheugh, shooch, sough = a trench, ditch, gutter, furrow, related to northern Middle English *sogh*, a swamp, rather like the Scots word *sauch* or saugh for willows, as in Sauchiehall in Glasgow, the willow hollow. Sheuch is used in both Northern Ireland and Scotland and in fact the Irish Sea / North Channel is referred to as 'The Sheuch' in Ulster.

Settlement Names

Biggin, bigging = building, cottage, hamlet, e.g. Newbigging, Lower and Upper Bigging(s), and the cottage in which Robert Burns was born at Alloway is called 'the auld cley biggin'.

Botl / bothl = house, dwelling, generally a high-status place, as in Newbattle, Morebattle (from OE *mere*, a lake) and Maybole in Ayrshire which is probably a maiden's house from OE *maege*, a maiden. However, it might be from Gaelic *magh*, a plain, so a house on the plain, maybe supported by earlier references to Meibothelbeg and Meibothelmor, with the Gaelic *beag*, little and *mòr*, big, added. It could even mean field of danger, from Gaelic *baoghal* (pron. buall), maybe a place to avoid, like the nearby Tarbolton which has perhaps added a Gaelic word (*tarbh*, a bull) to an English name. Like Bolton in England and Bolton in East Lothian, they were all once a farmhouse or village (*botl* and *tun*), though the latter still is.

Brig, brigg = bridge, from OE *brycg* (*bryggja* or *bru(r)* in Norse), but can sometimes refer to flagstones or reefs, which were maybe crossing places, e.g. the Auld and New Brigs o Ayr, Brig o Doon, Don or Dee, Briggait, Brighoose / house, Brigstanes or Stonebriggs, Brigton / Bridgeton or

Brigend / Bridgend of many towns or Stockbridge, meaning bridge of the tree trunks, from OE *stocc*, a tree. Many have, of course, been anglified to bridge, as in Drybridge, as opposed to Drybrig, a process that is rather ironic since 'brig' is much closer to the original old English as opposed to the modern English 'bridge'.

Burgh = town (from ON / OE *burg / burh*), a fortified dwelling (derived from the Latin *burgus*, fort or fortified town) as in Edinburgh, which does not mean Edwin's burgh, but probably town on the slope or rock, from Brittonic / Cumbric *eiddyn* (related to Gaelic *aodann*, a face) meaning slope or rock face, while Jedburgh is exactly what it says, town by the Jed Water (possibly from old Cumbric *gweden*, winding) and Roxburgh is from the personal name Hroc (rook). It was the setting up of royal and baronial burghs, which were granted special trading rights, that brought many Northumbrians to Scotland in the twelfth and thirteenth centuries.

Byre = a dwelling (originally), from the same root as the Norse *byr* (farm or hamlet) and related to the Norse word *bur* meaning hut or cottage. Eventually it just came to mean the cooshed / cowshed. Like the Norse *byr*, it is often shortened to *by*, as in various Humbies, or dog farms, from OE / Norse *hund*, a hound / dog.

Cir(i)ce / kirk = church, like the Norse *kirkja*, German *kirche* and Dutch *kerk*. In a land full of sinners, every town once needed several, though some have now been converted into pubs, probably leaving former ministers, or even the saints the churches were named after, birling in their graves! The list is very long indeed but here are a chosen few: Kirkcudbright, St Cuthbert's church / kirk, Kirkcolm, St Columba's Church, Kirkconnel, St Connel's Church, Kirkgunzeon, Guinnean / Finnian's Church, Kirkmichael, St Michael's Church, Kirkoswald, St Oswald's Church and

Kirkpatrick, St Patrick's Church. **Kirk** sometimes replaces the Brittonic **caer**, as in Kirkcaldy, meaning fort on the hard hill, from *caer caled din* or Kirkintilloch, hill-head fort, from *caer* and the Gaelic *cinn tulaich*. **Kirk** also replaces the Gaelic **kil**, though parallel forms (two versions that come from the same source) can be seen in Kilmichael and Kirkmichael, Kilconnel and Kirkconnel, with some following the *kirk* road rather than the *kil* road. Note how, in the above forms, Gaelic word order is used, i.e. the generic term (kirk) comes first and the qualifying or describing one second, e.g. the kirk of St Columba, but the following use normal Scots and English word order where kirk comes first because in these cases it is the qualifying term and not the generic one, as in Kirkden, church hollow or valley and not valley church, Kirkhill, church hill, Kirkton, church farm.

Cot = a small house, humble dwelling, sheep house, as in various cot / cote / coats names, such as Saltcoats, Coatdyke or Coatbridge. A cotter was a farm worker who lived in a small house, as in Robert Burns's poem 'The Cotter's Saturday Night'.

Erne = a house, e.g. Dreghorn, a dry house, or Whithorn, a white house.

Fauld, fald (OE *falud*) = enclosure used for cultivation or animals, small field, part of the outfield, used for grazing, as in Edgefauld, Langfauld, Lochfauld, Muirfaulds, Wheatiefaulds or Whitfaulds (*whit*, white) and many others.

Ha / haw = a large residence or farmhouse belonging to a wealthy landowner (the opposite of a cot), as in many ha / haw / hall names, from OE *heal* (or ON *holl*), e.g. Carterha, Gallowha, Thornyhaw or Hailes, though sometimes locals gave places ironic *haw* names to poke fun at the owners, as in Muttonhall and Cabbagehall in Fife.

Hag(g) / hags = enclosure, portion of woodland marked off for cutting, from OE *haga*, a hedge, enclosure, but it also meant the haw (berry) in hawthorn, from which we get Hagthorn places, such as the village of Haggs near Falkirk and Haggs Castle, Glasgow. **Haggs** (from Norse and OE) can also refer to **moss hags**, a hillock or muddy hollow, or slough, where peat had been cut in a bog, a name that features in the novel *The Men of the Moss-Hags* by S. R. Crockett and the short story *The Brownie of the Black Haggs* by James Hogg.

Haining (from ON *hegning*) = fence, hedge, wall, boundary or enclosure, often a protection, as in The Haining and many Haining places like Haining Brae, Haining Moss. To haine or hane something in Scots is to save or protect it.

Ham = homestead, village, hamlet. It is among the oldest Anglian words found in Scotland (mainly Lothians and Borders), often preceded by *ing*, and is often named after the owner or location, as in Coldingham which didn't necessarily lack heating as it was the village of the people of Colud (though they were probably cauld in winter), but Whittingham was the village of Hwita's people and, so that everybody knew where they were, Hampden was the village in the hollow, Yetholm was the hamlet at the pass or gate (see **yett**) and Tynninghame was the village of the people by the River Tyne. Birnam, a bit further north, was the warrior's home or village, from *bjorn*, a warrior, where a moving wood led to Macbeth's downfall, at least according to William Shakespeare. Nottingham and Birmingham in England have long outgrown their village status. Not to be confused with the Norse word *hamn*, a harbour, though it was sometimes confused with or changed to *holm*, a haugh, as in Leitholm and Yetholm.

Hus = hoose, house, a common Germanic word, same as in Norse, and we have hundreds of hoose place-names in

Scotland, though many have been changed to the modern English house, as in Stenhousemuir (from OE *stan*, stone), Auldhoose, Newhoose or Crosshouse, though it is still called Crosshoose in Ayrshire, where Scots has remained closer to the original sound of the word, so *hoose / hous* could be thought of as being more 'correct' than *house*!

Loan / lone (s) or **loanin / loaning** (OE *lane*) = lane, a grassy strip leading to pastures, cattle track through arable land, a green for milking cattle, so a path, track or roadway and nothing to do with asking for a loan. It is found in many Loans or Loanings, such as Dobbie's Loan or Tinker's Loan, Loanburn, Loanhead or Loaninghead, Loanknowes, Loanfoot, Greenloaning. The Cumbric equivalent is *lon* while the Gaelic *lànaig* (or *lonaig* and *lonaidh*) is a wee narrow track, possibly as in Lenzie and Leny or Lenny. Loans often lead through or to a **lea** or **ley**, grassland or meadow (from OE *leah*) as in Burns's song 'The Lea Rig', or as in Skelmorlie, Scealdamer's meadow, and also Lasswade, from *leas*, meadow and OE *gewaed*, a ford.

Raw = a row, line of houses with common features, used in many raw or row names, such as Langraw, Burnraw, Dykeraw, Fisherrow, Furnace Raw, Torryraw (tarry row). As they were usually inhabited by mining or fishing families, they were not salubrious dwellings. See the section on coal mining names at the end.

Stane = a stone, from OE *stan*, used in settlement names for stone buildings, as in Belstone / Balstane, near Dalry, North Ayrshire, possibly from Gaelic *baile*, a village, plus *stan*. Stenhousemuir was the stane hoose / stone house on the muir, but *sten* names in the Northern Isles are from the ON cognate *stein*, stone, though Shettleston, and Baillieston are from OE *ton / tun*, a farm. Athelstaneford in East Lothian, supposedly the ford of King Athelstan, might be a double tautological name: a ford stone stone ford, from Gaelic *àth*, a ford, and *ail*, rock, with OE *stan*, both meaning stone, so a well-stoned place.

Stede = farm, village, dwelling, piece of land, a homestead, as in various steid / stead places, especially in the Lothians and Borders, a cognate of ON *bolstadr* and often used insteid! **Stow**, found in the Borders, is another Old English settlement, town or meeting place word.

Tun / ingtun = enclosure, homestead, settlement, though the *ing* bit is an even older word, *inga*, from the very earliest Anglian migrations, meaning of the people, or followers of someone, so often preceded by a personal name, including some early forms added to Gaelic or Norse names. *Tun* is one of the most common OE names and was used over a very long period, originally a place of high status but later just came to mean farmstead or village and usually appears as *ton* at the end of the name, as in Bishopton, the Bishop's farm or village, Bowling, possibly the village of Bolla's people (though it may be from Gaelic *bò*, cow's and *linn*, a pool), or Haddington, the homestead of Hadda's people, Houston, Hugo's village (so the American pronunciation 'Hewston' is more correct), Kirkton, kirk farm, Newton, new farm, Milton, mill farm, Renton, Regna's farm, while Shettleston is the house of Seadna's daughter, from Gaelic *inghine*, of a daughter, with *toune / ton* added much later, Stewarton, farm of the steward, Stevenston, Steven's farm, Symington, Simon's farmstead and Uddingston, the homestead of Oda's people. Most names ending in 'town' are usually much more recent creations or have changed the older *tun / ton / toune* to town, ironically to make them sound like 'proper' English. Some *ston* names are however from stane / stone, the rock, as in Brunstane, Borestane, Lonestone, Thirlstane, though confusion sometimes arises with *ston* changed to stone.

Wic, wick = encampment, settlement, place or farm, generally a low status place, appearing as wick and distinguishable from the Norse *vik* / *wick* in that most examples appear inland, on both sides of the border, as in Alnwick, a dwelling by the River Aln in Northumbria, Berwick, barley farm, from OE / Scots *bere*, barley, Borthwick, either home or wood farm, from OE *bord*, table, plank or wood, Darnick, hidden settlement, probably by woods, from OE *diernan* to hide (as in darning), or Fenwick, a marshy place (fens). Note that many of these places have dropped the 'w' sound in the suffix, unlike Prestwick which is on the coast and was a priest's farm or house, from OE *preost*, a priest. Hawick is just a hawthorn hedge settlement, from OE *haga*, a hedge, Hedderwick, a heather farm, while Wigtown (and Wigton across the border) probably mean the farm or village of Wicga, an Old English personal name, though they could just mean farmstead. In southern England it usually appears as Wich, as in Norwich, Ipswich, or Greenwich.

Worth is an older word for enclosure, farm, replaced early on by *ton*, and we only have a few in or around the Border area, such as Polworth, Paul's farm, so it is not common in Scotland, though *worth* has been changed to *ford* in Cessford and to *burh* / *burgh* in Jedburgh and in some cases to *wood*.

Yett / **gate**, from OE *geat* = a pass, way, gate or entrance, but often just came to mean the road, as in Broadyetts, Canongate, Gateside, Gatehead / heid, Yetholm, Wateryet.

Scots Farming Names and Words

As you would expect, OE / Anglic and Scots words feature a great deal in farm names or agricultural practices, though some farm names have been anglicised over recent centuries. The Scots word *bour* or *bower* (ON / OE *bur*, *bu*) is found in some names referring to a tenant who hires cattle, a *bouman*, as in Bowhouse (see Norse **bur**). In Gaelic *buar* means a herd of cattle and *buachar* is one of the words for dung (*buwch* in Cumbric and Welsh is a cow, as in the name Buchan). Another word that comes from Old Norse and English is a toft, tuft, taft, or tupt, originally a house, site or foundation but came to mean cultivated land, as in Tofts or Toffs places, where people might unfairly be thought of as a bit snobbish!

Many words just describe farm size, shares position, condition or rent value (see affixes in intro to OE / Scots above), as in *muckle / meikle* (big), *laich* (nether or low), *heich* (high) *yonder / yonner* (referring to the outer or unenclosed fields), while some farms are described by their girth or *girt*, from Old Norse *gjord*, circumference, as in Girthill. If they were weel-aff / prosperous they possibly lived on a Mains ferm (home farm of an estate) or a Haa / Haw (hall) but maybe they just had a Holmbyre in a Cauldcleuch (cold hollow) beside Windygates or a Windyett, or maybe up on a Ruchill (rough hill) or even just a Cleghorn or Cley Biggin (clay building) in a Cotton (cot toun or farm cottage).

If they were tenant farmers they might have occupied a Pennyland, Merkland or Merksworth, from the OE *worth / warth*, a field or enclosure, though it might refer to its value or worth (from *merk*, an old coin), while they would often store crops in the Meiklelaucht or laft (barn or loft) or in a Barnton / Berntoun (a storehouse for barley or other cereals) which they might also have stored in a girnel or garnel (granary), near the kilnend of the steading or fermtoun.

They wid aften hae to muck oot (from Norse *myki*, cow dung) the *byre* (ON and OE) or coo shed an cairt the *shairn* or *sharn* (as in Sharon Street or Charing Cross) tae the midden or dunghill whaur they forked it wi a *graip* (another Norse word), while they wid cut the *rigs* o a crop or crap wi a *heuk* (hook, a sickle) an had tae keep the *sheuch* (ditch) clear. Mony a herd wid've hud tae collect the kye (cows) fae the Loanin or Pasturehill or aften leuk in *howes* (hollows) an *knowes* (hillocks) or ahint *dykes* or at the Dykeheid, faildyke or fauld-dyke for their *yowes* (ewes) or *wadders* / *wedders* / *wethers* (male sheep), *hoggs* (young sheep) an *stirks* (young bullocks).

Afore winter they micht hae gaen up tae the Auldmuir, Hag's Brae or Haggis Knowe for fuel tae keep the fires burnin (*hagg*, broken ground or trenches in the muir for the cutting of peats, but also coppiced woodland for brushwood or firewood) or mibbie up tae Nitshill, no for nits, but nuts! Efter a day wrasslin or chauvin / tauvin awa in the mire (from Old Norse and English *myrr*, a boggy place) they wid've cam hame in the mirk (from the Norse *myrkr*, dark, black), fair forfochan / forfochtin, deid duin, wabbit an mibbie crabbit! Nae wunner there are ferms caad Hell and Purgatory in Orkney!

Scots Flora and Fauna

As in Gaelic, shrubs and trees often appears in place-names, such as:

> aik, aiken = oak, oaken, as in Aikenheid / head, Aikenshaw
> birk, birken = birch, as in Birkhall (anglified from haa / haw, not a hall but a haugh or meadow), Birkenshaw (birchwood), Birkentap
> bourtree = elder, as in Bourtreebush / buss, Bourtreehill
> breckan / brechan = bracken, e.g. Brackenhills
> broom / brume = shrub, as in Broomhill or Broomielaw
> sauch(s) = willow(s), e.g. Sauchiehall, Sauchieburn or Saughton and from another willow word we get Wishaw, willow wood (OE *withig*, willow)
> shaw / schaw = a copse or wood, as in Shawfield, while Wormit in Fife does not refer to somebody suffering from worms, but to the Scots wormwood and a plantation of such trees.

A weel set oot ferm micht hae been on a Gowanlea (daisy meadow) or Claverhaugh (clover meadow or slope), wi mibbie a Broomyknowe or Broompark nearhaun tae get broom for their broomsticks (brushes) or for tyin doon their thatched roof, while dry whins (gorse) coud be cut on a Whinnyknowe tae set the fire ablaze. Nae doot there wid be plenty o haw bushes or hawtrees (hawthorns) on the Hawhill tae keep burds, like the *yorlin*, the yellow hammer, an the *gowdspink*, a goldfinch, happy, while maist hooses had a rodden tree (rowan) at the door tae ward aff the evil ee. In a guid summer they coud gether blaeberries up on the Blaeberry Craigs or brammles on the Brammle Bank or Brumley Brae, the latter being a place near Elgin and name of a fiddle tune.

Animals lik the brock (from Old English *brocc*, a badger) hae left their mark as weel, as in Broxburn or Strathbrock, tho Ibrox is fae the aulder Gaelic *bruic* (badger), but Broughton is fae OE *broc*, a brook. An auld Scots name for a hare is *cutty*, as in Kittyshaw (hare wood), Dalry, Ayrshire, while we hae some foxy places wi Tod / Todd (fox) in the name, e.g. Todhills or Todtails (foxgloves), an nae doot there are still a few Puddocky Ponds here an there. Shepherds still hae tae keep an ee open for gleds (kites / buzzards), as in Gled Brae or hoodie craws, as in Crawhill, baith great enemies o whaups or curlews.

11. OTHER LANGUAGES

Latin, French and Dutch
While the Celtic, Old Norse and Anglic / Scots tongues provide nearly all our place-names, other European languages have also left their mark here and there, especially the Latin and French words which often passed into both Gaelic and Scots in various forms.

Even the names we use for ourselves, especially Albannaich, Scots and Caledonians, came from the Romans, though these may well be Roman adaptations of native names, as with the Hebrides or Orkney, also called Pomona, an astonishing name for these islands, considering that she is the Roman goddess of fruit trees! From the Roman occupation we still have a number of Roman Roads, e.g. in Bearsden or in Straiton (from Latin *strata*, a street, via OE *straet*), plus a very long wall which can still be seen in various places, running from coast to coast (the Antonine Wall), various Roman camps and forts, some even with Latin names such as Chesters and Whitchester (from Latin *castra*, camp, via OE *ceastra*, plus *hwit*, white) or Trimontium (three hills), all in the Borders.

The names Penpont, north of Dumfries, a Br *pen* or top place, and Pont, near Stair in Ayrshire, are evidence of bridge places that came from Latin via Brittonic (Br *bont / pont*, which it still is in Welsh, from Latin *pontis*, of a bridge). In fact Brittonic had absorbed at least two hundred Latin words during the period of the Roman occupation of Britain, including *porth* for a door or gate and *ffenester* for a window.

However, most of the **Latin** names we have are a legacy of church Latin, not the Roman occupation, e.g. the Gaelic word *cill*, a church, is from the Latin *cella*, a cell or church and another Latin word for church, *ecclesia*, is at the root of the Brittonic and the Gaelic words for church, *egles* and *eaglais* (pron. EHKleesh), both close to the French word, église. Most of our Eccles or Eglis names originate from Cumbric as discussed in the Brittonic section.

We also have a number of Grange places which refer to pre-Reformation barns for storing the produce belonging to an abbey, derived from Anglo-French *graunge* from Latin *granica*, granary and *granum*, grain. Most places also needed a quarrel / correll / coral, not because they liked having rows, but just needed a quarry, as in Quarrel Hill, South Ayrshire, or the village of Quarrelton, a stony place near Johnstone, from the Latin *quarrelia*, a stone quarry.

In addition, we also have spittle or spittal places that sound as if people there had some rather disgusting habits, such as Spittalhill, near Riccarton and Spittalside, near Tarbolton, but thankfully they were all once hospitals, from Gaelic *spideal* or Scots *spital* (OE *spitel*, German *spital*, from Latin *hospes* and *hospitalis*), a refuge, spital or charity hospital.

Ministers usually had a *glebe*, from Latin *gleba* (a clod, a portion of land), in addition to their stipend (*stipendium*, pay, salary), Scots law is still full of Latin terms, while several legal or municipal posts, such as advocate and provost which come from Latin (or Latin via old French), are found in the names of streets or lanes in several towns or cities, such as Advocates Close in Edinburgh.

The capital might also have at least two names of Italian origin, or Italian via French: firstly the Sciennes area of the city derives its name from the convent once located there, named after the Italian St Catherine of Siena, which is Scienne in French and pronounced *Sheenz* in Scots, while some think the Pleasance took its name from the convent of St Mary of Placentia which once occupied that site, but this origin seems to have been invented by the historian William Maitland in his 1753 History of Edinburgh series. It is in fact from the old Scots term a 'plesance', meaning a park or garden, a word that is of French origin.

Some **French** names may come from the period when Norman French was used by the nobility from the eleventh to the thirteenth century, but most come from the Auld Alliance between Scotland and France which lasted about three centuries between the thirteenth and sixteenth centuries and well beyond in some aspects of our culture, something we can still see in place-names, e.g. Melville Castle, possibly from Malleville / mal ville, a bad town in Normandy or Belses (bel assize), Beauly (beau lieu), Chatelherault (from the French town of that name and title which the Duke of Chatelherault awarded to James Hamilton, Earl of Arran, in 1549), near Hamilton. We also find an estate in Moray called Haudmont, from *haut mont*, high hill, or the name Mountfleurie, in Leven, Fife, though the 'Frenchified' Marchmont in Edinburgh is in fact from the Gaelic *marc*, a horse, and *monadh*, a hill.

Our towns are not short of Vennels or lanes, Ports for gateways (from *porte*, a door or gate), Rows for streets (from *rue*), Cul de Sacs (literally the bottom of the bag) or dead ends, causeys or causieways (from Old French *caucie*, a paved road) for pavements or roadways, *syvers* (from Old French *essavier*, a drainage channel) or *stanks* (from Old French *estanc*, a pond) for drains and gutters. We also still have a *tron* in many of our cities and towns, as in the Tron Kirk and Trongate, from the old French word *trone*, a public steelyard or weighing machine.

Scotland has many places supposedly named after something Mary Queen of Scots did or said, such as the apocryphal Glasgow tale about how Polmadie got its name: Mary's horse, Pol, stumbled as she fled from the Battle of Langside in 1568 and she supposedly said 'Pol may dee that I maun leeve!' (see Gaelic **pol**). Likewise, according to legend, Boshang in Glen Shira, Argyll, is where she exclaimed 'quell beau champ!' Actually it is probably from the Gaelic word *both*, a hut or shelter, and *seang*, slim, slender, so just a wee shed!

Strawfrank in Lanarkshire was once the strath of the French, while Burdiehouse (Bordeaux house) and Little France, near Edinburgh, recall the Scottish love of claret and maybe the French servants (of Mary Queen of Scots and her French mother, Mary de Guise) who lived there around the middle of the sixteenth century – though this village probably originates with French cloth workers who settled there in the seventeenth century and set up a French mill (as named in General Roy's map in 1753). Another Edinburgh name with a similar origin is Picardy Place, once nicknamed 'Little Picardy', which takes its name from the Huguenot refugees who settled there after fleeing from Picardy in France in the late seventeenth century and who brought their silk-weaving skills with them.

A district of Pollokshields in Glasgow, near where the Battle of Langside was fought, has many streets named after Mary, such as Fotheringay Road (the place in England where she was kept prisoner for many years and eventually executed) or Mariscat Road (from Marie Scot) or Dolphin Road (from Dauphin, the French crown Prince, her first husband, but changed by locals to Dolphin!). Yet Dolphinton in the Lothians has nothing to do with either dauphins or dolphins, as it was simply a place belonging to a man called Dolfinn, while Moulin in Perthshire is not as French as it sounds, as it is from the Gaelic *maoilinn*, a bare round hill. Coshieville in the same district is in fact from the Gaelic *cois a*, beside the, and *bhile*, thicket or clump of trees, though the spelling has been Frenchified!

Even our grammar displays a French legacy, as we can see in the Scots habit of referring to a place, object, or even a health condition, by using the definite article, *the*, in front of it, something that comes from the French use of the definite article *le*, *la* or *l'* in front of nouns. Scottish children go to *the* school or *the* academy and maybe later *the* university or *the* tech. They might be good at *the* Maths or *the* French, or suffer from *the* cold or *the* flu, not just a cold or flu, and we go up or doon the toon, not uptown or downtown. Likewise some toons / touns or villages are referred to with *the* preceding the place-name, such as The Broch, The Brosie, The Jewel, The Den, The Lang Toon, The Muckle Toon, while some districts or parts of towns and cities are also often referred to in the same way, such as the Gorbals, the Canongate, the Trongate, the Grey Mare's Tail, the Howe of the Mearns, the Rest and Be Thankful, the Dirrans, the Blair, the Lynn, the Kerse, the Loans, the Glebe and many more.

Flemish / Dutch words have also passed into our language from Scotland's many commercial and cultural links with Holland and Belgium, especially before 1707, such as cran, crune, callant, loun, dowp, doitit, dub, geck, howf, hunkers, gowf, pinkie, redd up, scone or surnames of immigrant traders, fisherman or tradesmen, especially weavers, like Bremner (someone from Brabant), Wyper (from Ypres) or Fleming, the latter being found in a number of Lowland place-names, such as Flemington, Fleminghill and Flemyland, a farm near Kilwinning, Ayrshire, a county which also has a Dutch House Roundabout. Some Flemish names, though, were brought here by Norman incomers.

12. MODERN NAMES – SOCIAL, POLITICAL AND CULTURAL ISSUES

In contrast to older names, modern English names (i.e. from the last two to three centuries) rarely describe the local landscape, but are usually much more literal or functional, while at the other end of the scale, they are often designed to convey status or prestige, sometimes rather pretentiously, especially when named after people or places associated with royalty, nobility, power or fame, such as Alexandria, Bettyhill, Barbaraville, Fort George, Helensburgh, Port Charlotte, Prince's Street, Mount Vernon, or the large number of hospitals named after royalty by officials possibly hoping for royal patronage, such as the various Victoria and Royal Infirmaries, the Royal Alexandria in Paisley or the new Queen Elizabeth Hospital in Glasgow.

From the eighteenth century onwards, it became very fashionable to name places after royalty or nobility, firstly, to emphasise the power and prestige of the Union, or, more precisely, the House of Hanover in their various battles to overcome the Jacobite threat, and, secondly, to display the loyal credentials of a place, often removing the original Gaelic names in the process, e.g. Fort William (named after King Wiliam III by General Mackay in 1690), Fort George (named after George III) and Fort Augustus (named by General Wade after William Augustus, Duke of Cumberland, better known in Scotland as 'Butcher Cumberland), or Strelitz in Perthshire which was named after Queen Charlotte of Strelitz in Germany, wife of George III.

Edinburgh's eighteenth-century New Town (soon copied by other places) started following the Parisian fashion for naming streets after royalty, though it was actually George III who requested that the proposed St Giles Street be renamed Prince's

Street after his son, the Duke of York, the future George IV, the monarch who would make a famous royal tour of Edinburgh in 1822. Many other royal street names soon followed, such as Hanover Street, York Place, Charlotte Square, King George IV Bridge, George Street, though George Square was named after the architect's brother who was just plain George Brown. Not to be outdone, Glasgow also soon built its own royal thoroughfares, such as Union Street, George Street, Hanover Street and Queen Street.

However, an enthusiasm for naming places after local or national politicians, or generals that the dominant party wished to commemorate, has continued unabated for several centuries and there is still no shortage of them. From the eighteenth and nineteenth centuries, we have several Dundas Streets that celebrate William Pitt the Younger's 'uncrowned king of Scotland', Henry Dundas, as well as various Wellington Streets that commemorate the victorious general (and later Duke) at the Battle of Waterloo. We also have Palmerston Place, named after Lord Palmerston, and several places named after William Gladstone, two of the most famous nineteenth-century politicians. In more recent times we find many places honouring twentieth-century politicians like Churchill Gardens, Bevan Gardens, Keir Hardie Drive, Wilson Wynd or John Smith Court, or the many local councillors who are probably now best remembered for the streets named after them.

In contrast to all these prominent men, women have been conspicuously absent from Scotland's place-names, apart from royalty, saints and a few noble ladies, though in recent times this has started to change, e.g. Stevenston in Ayrshire has Mary Love Place in memory of a long-serving local pharmacist, as well as a Bonnie Lesley Court and Lesley Place, both in memory of a local girl, Lesley Bailie, whom Burns celebrated in song, while Saltcoats has a Betsy Miller Wynd in memory of the town's famous Victorian sea captain. Dalry now has a sheltered housing complex named after Bessie Dunlop, the sixteenth-century Dalry healer who was burnt as a witch in Edinburgh in 1576. Naming a place after a woman burnt as a 'witch' would have been unthinkable a few generations back.

Yet there is nothing new about all this, as forts, settlements or villages have often been named after the owner or some important local personage, or someone regarded as powerful or holy who has acquired varying degrees of status over time, e.g. all the places named after chiefs, monarchs or saints, but sometimes maybe just named after an ordinary person who left his or her name on the place he built or established, such as Winchburgh in West Lothian which is not where a lot of young folk did their 'winching' or where the lads went for a girlfriend (a wench), but is simply named after a man called Winca who was probably very proud of his fortified settlement or farm.

As well as political dimensions, names often have resonances of various kinds and inevitably certain names acquire positive or negative associations, from the status of the owner, the sort of people who live there and the degree of wealth or poverty of the area, or from the history of the place, its meaning, connotations and even the sound of the name or its use in literature, music and commerce.

There is of course an influential social dimension or status attached to certain parts of any town or city, as some affluent places have come to be thought of as 'posh', or were once thought of as posh, such as Kelvinside in Glasgow or Morningside in Edinburgh, while other places may have seen better days, or maybe have acquired a rough or poor reputation, though this may also change with time. I have already mentioned in the introduction that a lot of names were anglified in the eighteenth and nineteenth centuries to give them greater status, as many older Scots or Gaelic ones were thought to be too common or even vulgar, such as the village of Muttonhole in Edinburgh which was changed to Davidson's Mains, while Victorian sensibility certainly could not cope with the original name of the famous Storr rock pinnacle on the Isle of Skye which was renamed *bodach* (old man) instead of *bod* (penis) as the Old Man of Storr was thought to be much more respectable.

Today we often find new houses or streets named after Scottish rivers, mountains, islands, castles, or even famous golf courses (perhaps with a local connection, though often not), no doubt chosen because of their historical significance, legendary, romantic or merely sentimental personal associations. Thus we often find Highland names which have wandered far from their original location, such as Lochnagar, Loch Ness, Loch Lomond, Fingal's Cave (also the name of Mendelssohn's famous piece of music), the Trossachs, Glenfinnan (with Bonnie Prince Charlie and Harry Potter associations), while many of our islands have also drifted far from their moorings, such as Iona, Staffa, Barra, the Isle of Mull or Isle of Skye, whose bens, glens, bays and townships often feature in Lowland place-names or even in place-names across the world. (See below.)

Songs or poems have of course immortalised many such places, particularly the islands, while many places have been given names associated with our most celebrated writers, especially Robert Burns, such as Alloway, Mauchline, Mossgiel, Lochlea, Ellisland, Doon or Afton. In fact our national bard features in over 470 road names across the UK, including nineteen in London, though Glasgow has the highest concentration with seventy-two streets featuring his name and, not surprisingly, Ayr second with twenty-five. Dalry in Ayrshire even has a street named after the town's Burns Club, the oldest continuous Burns club in the world.

However, Sir Walter Scott is maybe not far behind, as there was a great fashion in the nineteenth and early twentieth centuries for names related to him, such as Abbotsford, Hazeldean or Lochinvar, featuring in houses or streets, or even railway engines, stations (Waverley) and paddle steamers, e.g. the *Lucy Ashton*, the *Jeannie Deans* and again the *Waverley*. Parts of both Edinburgh and Glasgow have places named after his titles, characters or places in his novels, such as Knightswood in Glasgow where there are streets named after characters from *Ivanhoe* while other Scott novels or characters are commemorated in places such as Ravenswood Road, Dinmont Road (both Shawlands) and Peveril Avenue (Rutherglen). Scott's contemporary writer, John Galt, is only commemorated with sheltered housing named after him in Greenock, as is Robert W. Service in Kilwinning, while Irvine used to have a John Galt Primary School, but still has a Galt Avenue and a rather shady old close, named, perhaps not inappropriately, after one of his most famous characters, Provost Pawkie.

In more recent times, as a tribute to another of our capital's famous literary sons, streets have been named after places and people from the novels of Robert Louis Stevenson, with Balfour Loan, Duart Crescent, Alan Breck Gardens and Hoseasons Gardens, though, unsurprisingly, there are no Dr Jekyll and Mr Hyde Gardens, in spite of the fact that the novel was perhaps inspired by an infamous Edinburgh character, Deacon Brodie, and Edinburgh was very much on his mind when he wrote it. Of course, Jekyll and Hyde Gardens might harm the city's image or attract the wrong kind of residents, though there is a Jekyll and Hyde pub which no doubt attracts an intriguing sort of clientele!

Not to be outdone in poetic status, Glasgow has streets named after the English novelist Charles Dickens and poets Shelley, Byron and Shakespeare, though, sadly, it does not seem to have been very enthusiastic about naming places after its own literary sons or daughters. Many Glasgow streets were, however, named after wealthy businessmen, entrepreneurs and industrialists during the eighteenth and nineteenth centuries, something that is not always apparent, as in the curious case of William Harley, a textile manufacturer and builder, who is commemorated via Bath Street because he built a series of public baths along this street which led to 'Willow Bank Pleasure Gardens' on his estate at the northern end of what is now one of the city's main arteries.

We also have a long history of folk familiarisation, or popular-naming alternatives to official names, especially using colloquial diminutives or nicknames, to express affection, humour or ridicule, such as Auld Reekie (even though it originally came from the Gaelic *alt-ruighe*, a high slope), Paddy's Milestone or Paddy's Market, the Heilanman's Umbrella, the Dough School, the Squinty Bridge, the Armadillo, Paradise, Porty, the Lang Toon, Fittie (i.e. Footdee in Aberdeen, though it has nothing to do with feet, as like Fitty Loch and Foodie Hill in Fife, they are peaty or turfy places, from Gaelic *foid*), Drummy, Feegie Park, a Weegie, the Techy, the Yoony, the Cally, the Vicky, the Queen Betty Hospital! The name Dumbiedykes or Dummydykes in Edinburgh would certainly not be acceptable today, as it actually comes from a school for the deaf and dumb which was set up there in the eighteenth century and became known as the 'dummy hoose', with the dykes added later.

Scotland also abounds in quirky or amusing names, as discussed earlier, like the Puddliedoodly in Irvine, a name of debateable origins, while we also find places quaintly named after a memorable character or contemporary event, or in one case at least, a memorable animal: a famous nineteenth-century racehorse called Beeswing, which once had an inn named after it and the name of the inn eventually became the name of the surrounding village in Dumfries and Galloway. In contrast to a buzzing village, you will find a long steep climb on the

Arrochar to Inveraray road (the A83), in Argyll, called the 'Rest and Be Thankful', a name given to it by the labourers building the old military road in 1753 who were no doubt thankful to reach the top of that very steep hill. On the other side of the 'Rest' you will find, near Butter Bridge, a farm road leading up to a house called 'Abyssinia' (what is now Ethiopia), apparently named after an old soldier who either worked building the place or lived there as a shepherd and who was renowned for telling everyone about his military adventures in Abyssinia!

We can also find various other exotic foreign imports on the map, as well as quite a few biblical names, such as Padanaram, Jericho, Joppa (the modern Jaffa in Israel), Jordanhill or Jordan Burn, some of them no doubt reflecting the religious devotion or piety of their owners, for, as David Ross comments in *Scottish Place-names*, 'biblical names are not uncommon in Scotland from the sixteenth century onwards, though usually applied to farms, or even fields, rather than large communities'. Students could perhaps investigate some of these in their own area.

This leads us into the domain of house names, the last refuge of private naming, as opposed to names chosen by officialdom. While house and property names might just reveal interesting or idiosyncratic personal choices, such names often tell us a lot about the culture of the time they come from, whether it is a display of religious zeal, historical or geographical fascinations, fluctuating fashions or even literary and musical passions. At the top end of the social scale we have, as with street names, no shortage of choices intended to bestow status and style, as can be easily seen by walking around any of the more affluent parts of our towns and cities where rather pretentious foreign-sounding names are not uncommon, like Bellevue, Belmont, Claremont, or Jemimaville.

Due to cheap foreign holidays, more recent decades have also witnessed a plethora of Spanish or Italian house names like Casa Bona, Casa Grande or Casa Blanca and even a few La Manchas, a place made famous by the Spanish novel (but probably the film version) of *Don Quixote* by Miguel de Cervantes, whose hero came from La Mancha, though this famous Spanish name was first bestowed on a manor house in Peebleshire by a retired eighteenth-century admiral who had spent some time in La Mancha.

Yet, just as we have streets named after famous Scottish islands and mountains, we also seem to have a great love of naming houses after famous or favourite Scottish places or just places with personal nostalgic or family connections, while there are even many houses named after other famous Scottish homes, like Abbotsford or Balmoral, or the names of Robert Burns's farms, like Lochlea and Mossgiel.

Likewise, we have the quite remarkable phenomenon of Gaelic house names, even in non-Gaelic-speaking areas, sometimes as simple as Mo Dhachaidh / 'Ar Dachaidh' (my / our home), Taigh na Mara (sea house), or Taigh Solais (lighthouse or house of happiness or comfort). These names have often been adopted by people who have no Gaelic, or only a distant Gaelic ancestor, though it may again reflect a love of Scottish places, especially islands, or possibly a pride in displaying a uniquely Scottish name, possibly a nostalgia for a lost Gaelic heritage or a desire to reclaim it. There is even a medical practice in East Ayrshire which has adopted the old Gaelic (or even Cumbric) name for the town of Drongan (itself a Gaelic name), Taiglum, which possibly means the house by the gorge or chasm, from Gaelic *taigh* and *glom* (abyss or chasm).

Pubs have also long held on to, or even revived, Gaelic or Scots names, perhaps to suggest longevity, ethnicity, or even authenticity, using words like 'Slàinte' or 'Fàilte', or simply 'The Auld Hoose', or (at the time of writing) 'Drouthy Neebors' pub in Edinburgh, the Gallus Bar in Glasgow, and a cafe in Dalry, Ayrshire, called 'The Twa Bees', presumably to suggest how hard the owners work. I have even noticed a revival in using Scots for house names, not just preserving old local place-names, but simply using Scots words as opposed to English ones, perhaps reflecting a desire to display or even proclaim the status and uniqueness of the owner's 'mither tongue', with names like 'Oor Hoose', or 'Tru Hame', while I have even come across a 'But 'n Ben'.

13. WORK AND PLACE-NAMES

In contrast to topographical and ownership names, many places (and people) are simply named for the work done in a place or the occupations found there, such as the many old farm-related names still found in our towns and cities, like Barns places in Ayr, Byres Road in Kilwinning and Glasgow, Sharon Street in Dalry which means the same as Glasgow and London's Charing Crosses, i.e. they were all once places covered in coo dung, thus reminding us how close our old towns and cities once were to the surrounding countryside.

Yet in the middle of these old towns we also find wynds, lanes and closes recalling the many former trades or professions once located there, such as the labyrinth of closes in Edinburgh's old town named after various trades or crafts, such as Baxter's Close, Skinner's Close, Old Fishmarket Close, Fleshmarket Close, though many closes and wynds were simply named after the owners, chief resident or occupiers, such as Hastie's Close, Milne's Court, Paterson's Court or Blackfriars Wynd.

Likewise, many areas or towns were named after the main local industry, such as Saltcoats, Prestonpans (both from salt-making), Potter Row, Edinburgh, or Potterhill, Ayrshire, Wauk Mill or Fullerton (both from waulking or fulling, i.e. cleansing and thickening cloth), Mill Road or Furnace Raw, while many people took their name from their occupation, e.g. Gow (blacksmith), Caird (craftsman), or simply took their names from the place they lived, such as Leslie, Lennox, Urquhart. An old Scottish custom was also to refer to someone by the land they owned or the farm they worked, such as Cameron of Lochiel, Blair of Blair, Kerr of Kersland.

Just as farms or villages were often named after their owners, industrial sites or facilities were often just as proprietorial in their naming, especially mines or mills which often simply took their owner's name, such as Biggart's, Kyle's or Knox's Mills in the Garnock Valley, though mines were sometimes merely given numbers, leaving behind ruins or bings, long known as 'Nummer Seivin' or whatever. However, they also often took their name from an older farm or landscape feature, such as the Briery Sink, Todhill, the Reddance, Pitcon (a name that predates the pit), all in the Garnock Valley, while in the Lothians we have Bonnyrigg, Blinkbonny and 'The Brosie', a local name for the Gilmerton Colliery on the outskirts of Edinburgh, while the Roslin Colliery was known as 'The Moat' after a local farm.

Perhaps the widespread use of local or familiar names for the place of work, especially collieries, suggests a need for folk to put their own stamp on a place, almost to claim it as their own, often revealing a pride in their community. In some places there was even maybe a feeling among some miners that, although they didn't own the mines, the mines did belong to them in some way, while in a deeper sense they also belonged to the mines, as they toiled, sweated, bled and sometimes died in them, as in the 'Auld Barny' (Bartonholm No. 3 Pit), Irvine, where a fatal explosion took place in 1871.

All mining communities possessed a powerful sense of identity, brotherhood and community spirit, rooted in a shared set of moral values that were surely born of an absolute dependence on workmates in defending each other from the hazards of the mine, or from an exploitative employer. Folk in miner's raws really didn't have to lock their doors at night, as no one would ever think of stealing from their neighbours and in any case, they had little that was worth stealing. A mining community that certainly displayed this strong community bond was Glenbuck in South Ayrshire (long gone but commemorated by a monument) which produced an astonishingly successful football tradition built around their village team, the famous

Glenbuck Cherrypickers who no doubt picked up plenty coal and a number of trophies, but gey few cherries.

Another dimension to this process of folk familiarisation, or community name claiming, was that collieries were often given by-names / nicknames, while even sections or areas within collieries were similarly named, sometimes humorously or ironically, such as the Ponderosa (the name of a Nevada ranch in a popular 1960s television Western called *Bonanza*) and the Khyber Pass (a much fought over mountain pass between Pakistan and Afghanistan) at Bilston Glen in Midlothian, while in contrast, the Boglemart Pit in Stevenston, Ayrshire, was known locally as the Jenny Lind pit, named after the famous nineteenth-century Swedish soprano, maybe because some of the miners were fine singers.

Conversely, types of coal, coal seams, mining techniques or functions often bequeathed names like the Loading Bank in the Garnock Valley. However, the former mining village of Shotts in North Lanarkshire is not derived, as is sometimes thought, from the use of explosive shots in the pits, but from an Old English word *sceots*, meaning steep slopes. Yet there is no doubt that in The Jewel, in Niddrie, near Edinburgh, we have a real gem of a name, as it is from a type of coal called jewel, while other street names in the area are also named after coal seams, such as Peacock, Blackchapel and Vexhim, some of these names used ironically.

Some local names also relate to events around the time the colliery was sunk or opened, such as the Klondike in Newcraighall, East Lothian, a name imported from the Yukon Goldrush at the end of the nineteenth century, possibly to describe a pit where a lot of money could have been made at that time, especially by the owners. Or perhaps a lot of miners, who failed to find gold, returned to dig black diamonds there? Similarly, Polkemmet in West Lothian was known as the Dardanelles, perhaps because conditions in this pit reminded miners of this disastrous military campaign during the First World War, probably named by former soldiers who returned from the war to work there.

Scotland's central industrial belt, from Ayrshire and Lanarkshire to Fife, was until around the middle of the twentieth century still covered in hundreds of miners' raws / rows which were not exactly commodious residences, often just with one room or one bedroom and kitchen, usually without inside toilets or running water, though water sometimes ran down the walls. These raws all had names, sometimes from an older place-name, e.g. Peesweep Raw or the Dandy Raw (better-quality houses for foremen) in Dalry, Ayrshire, or from owners' names, such as Cowan's Raw in Hurlford, or just from a site position, location or size, such as Front, Back, or Lang Raw, or proximity to a topographical feature like Burn Raw or Schuil Raw, or an industrial site or function, such as Furnace Raw, or Washer Raw.

Another curious feature is that many of the miner's raws were also given nicknames, sometimes ironic, to describe the type of accommodation or the type of people who occupied it, often perhaps an expression of a quirky sense of humour, e.g. Cadgers' Raw in Hurlford, Monkey Raw in Glengarnock, Stickit (unfinished) Raw in Dalry, or Honeymoon Raw in New Cumnock where newly married couples often set up home.

Other industries or products have long used Scottish names to promote a distinctively Scottish identity as an important selling point, e.g. using alternative Scottish names like Atholl, Albion or Caledonian, or river and mountain names, while modern industrial or corporate names are often designed to impress or create an image of prestige, power, bravery and strength in similar ways, by using names for oil fields like Auk, Brent, Beauly, Buchan, Caledonia, Claymore, Iona, Ivanhoe, Highlander, Piper, Rob Roy and even Scotty and Tartan!

14. SCOTTISH NAMES WORLDWIDE

Over many centuries, Scots have left their native land, either voluntarily or otherwise, in very large numbers and it has been estimated that more than twenty million people worldwide have Scottish ancestry. Due to this diaspora, Scottish names have been taken worldwide, especially to North America and Australasia where we find multiple examples of places named after somewhere in Scotland, e.g. there are so many Aberdeen places throughout the world that a book has been written about them. Indeed, it is quite astonishing to discover just how many places there are throughout the world named after somewhere in Scotland.

There are hundreds of Tay, Tweed, Spey and Clyde names worldwide, especially Clyde, and there are at least 550 places in South Africa alone that have Scottish names, including a Tilliedudlem near Durban. Thousands of Canadian and American towns or cities have been named after a Scottish place, like Albany, Calgary, Banff, Dallas, Hamilton, Houston, and there are more than two hundred localities with Scottish names in the New York Metropolitan area alone, while even in the Rocky Mountains in Canada you will find a place where the railway lines from east and west first met called Craigellachie!

Thus a very high percentage of North American or Australasian names are of Scottish origin, including many in Gaelic, especially in Nova Scotia and around Dunedin in New Zealand. In fact, there is roughly a ratio of one Scottish name to every four of English origin in Australasia, South Africa, as well as across North America and the Caribbean, especially in Jamaica which has at least twenty-five Scottish place-names, including a Kilmarnoch (Kilmarnock), and you will find an Arthur's Seat on Barbados. Hong Kong has over thirty Scottish place-names and even Antarctica has about forty, including an Ailsa Craig! Canada has its Nova Scotia, France has its New Caledonia, but Argentina has a Nueva Escocia (New Scotland). Chile has a place called Cochrane, named after the Scottish Admiral who helped liberate their country from the Spanish, and they also have an Alejandro Selkirk Island, named after the famous castaway, Alexander Selkirk, whose adventures inspired Daniel Defoe's novel *Robinson Crusoe*.

But in our space age, Scottish names aren't just limited to planet Earth, as, believe it or not, our place-names now zoom out into the wider universe, including places where no man has set foot. Some planets, and even places further into outer space, are now illuminated with names from somewhere in Scotland, including Martian craters called Ayr, Darvel, Doon, Balvicar, Banff and Echt, while elsewhere on the Red Planet you will find a cliff called the Maria Gordon Notch, which Nasa named after the pioneering Scottish geologist, Dame Maria Ogilvie-Gordon. You will also find, if you ever go there, an Arran Chaos, Callanish and Torrisdale on Europa, one of Jupiter's moons, but you should also look out for asteroids whizzing past called Clackmannan and Fingal, as in Fingal's Cave on Staffa.

Many places throughout the world are also named after Scots who left their mark in their new country, whether as explorers, governors, soldiers, politicians or businessmen, including Admiral Cochrane, mentioned above, or Andrew Carnegie, one of the wealthiest men on the planet around the turn of the twentieth century, a ruthless industrialist, but also a philanthropist who gave away most of his money and left libraries named after him worldwide. Hundreds of other places are named after famous Scots, like Burns, Scott, Wallace or Livingstone, e.g. there are Aftons (from the Burns song) in Iowa, New York State, Minnesota, Oklahoma, Wyoming and an Afton Canyon in California, while there is a community (originally a separate town) in Ontario called Galt, named after John Galt, the Irvine-born writer who also founded the city of Guelph in Ontario.

Likewise, certain iconic Scottish brands or products have travelled far and become much sought after, mainly due to their unique qualities, such as Ayrshire cows, Aberdeen Angus cattle, and of course steaks, Paisley patterns, Harris Tweed, Crombie coats, Fair Isle sweaters, all of which are renowned global names. Undoubtedly, many Scottish place-names are known and much loved worldwide because of the fact that many varieties of 'Scotch' (whisky) carry our place-names on their labels!

Conversely, Scots who have travelled the world have also brought back foreign names or had places named after the battles they took part in, especially the Battle of Waterloo with around a dozen Waterloo places in Scotland, including Skye and Prestwick and there is even a Waterloo Point on the Wee Cumbrae, but we also find plenty other battles commemorated in our place-names such as The Alma (Arbroath, and Brodick on Arran), Camperdown (Dundee), Inkerman (Ayr), Ladysmith (Kilbirnie), Omoa (Lanarkshire), Portobello (Edinburgh), while emigrants to the USA, or maybe returning emigrants, are remembered in names like California and the Klondyke in Stirlingshire, both named after gold rushes, or Dakota Place in Dalry, Ayrshire.

Some of these names could provide interesting local research for students. An Ayrshire mining village is named after Patna in India where a local landowner, Provost William Fullerton, was born and where his father made his fortune in the eighteenth century, while Mount Vernon in Glasgow was given this name by a wealthy eighteenth-century tobacco lord, George Buchanan, who thought Windyedge was much too common and renamed it after George Washington's plantation in Virginia.

Virginia should also remind us that much of Glasgow's eighteenth-century wealth was derived from tobacco, tea, sugar or cotton, and some of our wealthiest tobacco lords are remembered in names like Ingram, Buchanan or Glassford Street, while Jamaica, Tobago and Virginia Streets and Plantation Quay remind us of places where Scottish landowners and merchants were heavily involved in various aspects of the inhumane but highly profitable slave trade, something that helped fuel our industrial revolution, and a topic that has recently aroused much controversy, with some people campaigning to have these names changed.

In contrast to this unhappy episode in our history, immigration or town twinnings have, in more recent times, gifted us names like Barga Gardens, in Saltcoats, which celebrates the many Italian Scots originating from that small northern Italian town who made Scotland their home and who cherish the many positive links that happily now exist between both places.

Students could be asked to research some of the above foreign influences or dimensions or explore particular people, activities, battles or events which left a mark in their own area.

15. CONCLUSION

> Place-names have come a long way and most of them have gathered resonances.
>
> —Professor Bill Nicolaisen

In this guide I have tried to provide some knowledge that should at least help readers to unfankle Scotland's place-names and hopefully make them eager to learn about the places around them. As we have seen, Scotland has an exceptionally rich and diverse linguistic heritage for a small country, possibly more diverse than any country of a similar size and even more than many larger ones. Surely this is something we should cherish and ensure our young folk appreciate. If you are a teacher, I hope you will consider trying to interest your pupils in their local place-names as it will take them on a fascinating journey, discovering so many interesting things about where they live, hopefully a journey which can be long and winding and maybe never-ending.

16. BIBLIOGRAPHY

Allan, Elizabeth, and Adam Watson, *The Place Names of Upper Deeside* (Aberdeen University Press, 1984), and other more recent books by Watson, e.g. *Place-names in much of north-east Scotland* (Paragon, 2013).

Crehan, Elfreda, and Peter Terrell, *What's in a Scottish Place-name?* (Lexus Ltd, 2016).

Dorward, David, *Scotland's Place-Names* (Blackwood, 1976).

Drummond, Peter, *Scottish Hill Names, their Origin and Meaning* (Scottish Mountaneering Trust Publications, 2007).

Eyers, A. M., *Scottish Place-Names* (Celtic Publications, 1980).

Field, John, *Place-Names of Great Britain and Ireland* (David & Charles, 1980).

Johnston, J. B., *Place-Names of Scotland* (S. R. Publications, 1979).

Johnstone, Fiona, *Introducing Scotland in Place-Names* (Spur Books, 1982).

Mackay, George, *Scottish Place-names* (Geddes & Grosset, 2007; Waverley Scottish Classics, 2009).

Markus, Gilbert, *The Place-names of Bute* (Paul Watkins, 2012).

Markus, Gilbert and Simon Taylor, *The Place-names of Fife*, 5 vols (Shaun Tyas, 2011), and 'Scotland and the Flemish People', University of St Andrews, an online article drawn from Peadar Morgan's PhD thesis (both very scholarly).

McNeill, P. and R. Nicholson (eds), *A Historical Atlas of Scotland* (St Andrews, 1975).

Mills, A. D., *A Dictionary of British Place-names* (Oxford University Press, 2011).

Murray, John; Moireach, Iain, *Reading the Gaelic Landscape* (Whittles Publishing, 2016).

Nicolaisen, W. H. F., *Scottish Place-Names* (Batsford, 1976; John Donald, 2001)

Ross, David, *Scottish Place-names* (Birlinn, 2001).

Taylor, Iain, *Place-names of Scotland* (Birlinn, 2011).

Watson, W. J., *The History of the Celtic Place-names of Scotland* [1926] (repr. Birlinn, 1990 and 2011), is still the primary scholarly reference guide on the subject and has been christened the 'Old Testament' of Scottish place-names, while *Scottish Place-Names* by W. H. F. Nicolaisen has been christened the 'New Testament' of Scottish place-names.

Wood, N., *Scottish Place-Names* (Chambers, 1989).

Crehan and Terrell, Dorward, Eyers, F. Johnstone and Mackay are cheap and easy for teenagers to use. Field is a dictionary, while Watson, Nicolaisen, Murray and Taylor are very scholarly.

17. ONLINE RESOURCES

A wide array of useful resources can be found online, for example:

Ainmean-Aite na h-Alba / Gaelic Place-names of Scotland
www.ainmean-aite.scot

Dictionaries of the Scots Language
dsl.ac.uk

Education Scotland
education.gov.scot

Learn Gaelic online dictionary
www.learngaelic.scot

Ordnance Survey #GetOutside guides
getoutside.ordnancesurvey.co.uk

Scots Language Centre
www.scotslanguage.com

Scots Words and Place-Names
www.swap.nesc.gla.ac.uk

Scottish Place-Names Society
www.spns.org.uk

Understanding Scottish Places
www.usp.scot

Milton Keynes UK
Ingram Content Group UK Ltd.
UKHW022121180924
448454UK00010B/247